Universitext

T0202870

Anton Deitmar

A First Course in Harmonic Analysis

Second Edition

 Springer

Anton Deitmar
Department of Mathematics
University of Exeter
Exeter, Devon EX4 4QE
UK
a.h.j.deitmar@exeter.ac.uk

Mathematics Subject Classification (2000): 43-01, 42Axx, 22Bxx, 20Hxx

Library of Congress Cataloging-in-Publication Data
Deitmar, Anton.
    A first course in harmonic analysis / Anton Deitmar. – 2nd ed.
       p. cm. — (Universitext)
    Includes bibliographical references and index.
    ISBN 0-387-22837-3 (alk. paper)
    1. Harmonic analysis.   I. Title.
    QA403 .D44 2004
    515′.2433—dc22   2004056613

ISBN 0-387-22837-3   Printed on acid-free paper.

Photocomposed copy prepared from the author's LaTeX files.

Printed in the United States of America.   (MP)

9 8 7 6 5 4 3 2 1   SPIN 11019138

springeronline.com

Preface to the second edition

This book is intended as a primer in harmonic analysis at the upper undergraduate or early graduate level. All central concepts of harmonic analysis are introduced without too much technical overload. For example, the book is based entirely on the Riemann integral instead of the more demanding Lebesgue integral. Furthermore, all topological questions are dealt with purely in the context of metric spaces. It is quite surprising that this works. Indeed, it turns out that the central concepts of this beautiful and useful theory can be explained using very little technical background.

The first aim of this book is to give a lean introduction to Fourier analysis, leading up to the Poisson summation formula. The second aim is to make the reader aware of the fact that both principal incarnations of Fourier theory, the Fourier series and the Fourier transform, are special cases of a more general theory arising in the context of locally compact abelian groups. The third goal of this book is to introduce the reader to the techniques used in harmonic analysis of noncommutative groups. These techniques are explained in the context of matrix groups as a principal example.

The first part of the book deals with Fourier analysis. Chapter 1 features a basic treatment of the theory of Fourier series, culminating in L^2-completeness. In the second chapter this result is reformulated in terms of Hilbert spaces, the basic theory of which is presented there. Chapter 3 deals with the Fourier transform, centering on the inversion theorem and the Plancherel theorem, and combines the theory of the Fourier series and the Fourier transform in the most useful Poisson summation formula. Finally, distributions are introduced in chapter 4. Modern analysis is unthinkable without this concept that generalizes classical function spaces.

The second part of the book is devoted to the generalization of the concepts of Fourier analysis in the context of locally compact abelian groups, or LCA groups for short. In the introductory Chapter 5 the entire theory is developed in the elementary model case of a finite abelian group. The general setting is fixed in Chapter 6 by introducing the notion of LCA groups; a modest amount of topology enters at this stage. Chapter 7 deals with Pontryagin duality; the dual is shown to be an LCA group again, and the duality theorem is given.

The second part of the book concludes with Plancherel's theorem in Chapter 8. This theorem is a generalization of the completeness of the Fourier series, as well as of Plancherel's theorem for the real line.

The third part of the book is intended to provide the reader with a first impression of the world of non-commutative harmonic analysis. Chapter 9 introduces methods that are used in the analysis of matrix groups, such as the theory of the exponential series and Lie algebras. These methods are then applied in Chapter 10 to arrive at a classification of the representations of the group SU(2). In Chapter 11 we give the Peter-Weyl theorem, which generalizes the completeness of the Fourier series in the context of compact non-commutative groups and gives a decomposition of the regular representation as a direct sum of irreducibles. The theory of non-compact non-commutative groups is represented by the example of the Heisenberg group in Chapter 12. The regular representation in general decomposes as a direct integral rather than a direct sum. For the Heisenberg group this decomposition is given explicitly.

Acknowledgements: I thank Robert Burckel and Alexander Schmidt for their most useful comments on this book. I also thank Moshe Adrian, Mark Pavey, Jose Carlos Santos, and Masamichi Takesaki for pointing out errors in the first edition.

Exeter, June 2004   Anton Deitmar

Leitfaden

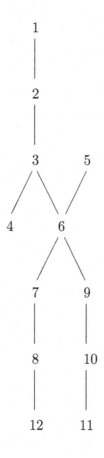

Notation We write $\mathbb{N} = \{1, 2, 3, \ldots\}$ for the set of natural numbers and $\mathbb{N}_0 = \{0, 1, 2, \ldots\}$ for the set of natural numbers extended by zero. The set of integers is denoted by $\mathbb{Z}$, set of rational numbers by $\mathbb{Q}$, and the sets of real and complex numbers by $\mathbb{R}$ and $\mathbb{C}$, respectively.

Contents

Part I

Fourier Analysis

Chapter 1

Fourier Series

The theory of Fourier series is concerned with the question of whether a given periodic function, such as the plot of a heartbeat or the signal of a radio pulsar, can be written as a sum of simple waves. A *simple wave* is described in mathematical terms as a function of the form $c\sin(2\pi k x)$ or $c\cos(2\pi k x)$ for an integer k and a real or complex number c.

The formula

$$e^{2\pi i x} = \cos 2\pi x + i \sin 2\pi x$$

shows that if a function f can be written as a sum of exponentials

$$f(x) = \sum_{k\in\mathbb{Z}} c_k e^{2\pi i k x},$$

for some constants c_k, then it also can be written as a sum of simple waves. This point of view has the advantage that it gives simpler formulas and is more suitable for generalization. Since the exponentials $e^{2\pi i k x}$ are complex-valued, it is therefore natural to consider complex-valued periodic functions.

1.1 Periodic Functions

A function $f : \mathbb{R} \to \mathbb{C}$ is called *periodic of period $L > 0$* if for every $x \in \mathbb{R}$,

$$f(x+L) = f(x).$$

5

If f is periodic of period L, then the function

$$F(x) \;=\; f(Lx)$$

is periodic of period 1. Moreover, since $f(x) = F(x/L)$, it suffices to consider periodic functions of period 1 only. For simplicity we will call such functions just *periodic*.

Examples. The functions $f(x) = \sin 2\pi x$, $f(x) = \cos 2\pi x$, and $f(x) = e^{2\pi i x}$ are periodic. Further, every given function on the half-open interval $[0, 1)$ can be extended to a periodic function in a unique way.

Recall the definition of an *inner product* $\langle .,.\rangle$ on a complex vector space V. This is a map from $V \times V$ to $\mathbb{C}$ satisfying

- for every $w \in V$ the map $v \mapsto \langle v, w\rangle$ is $\mathbb{C}$-linear,

- $\langle v, w\rangle = \overline{\langle w, v\rangle}$,

- $\langle .,.\rangle$ is positive definite, i.e., $\langle v, v\rangle \geq 0$; and $\langle v, v\rangle = 0$ implies $v = 0$.

If f and g are periodic, then so is $af + bg$ for $a, b \in \mathbb{C}$, so that the set of periodic functions forms a complex vector space. We will denote by $C(\mathbb{R}/\mathbb{Z})$ the linear subspace of all *continuous* periodic functions $f : \mathbb{R} \to \mathbb{C}$. For later use we also define $C^{\infty}(\mathbb{R}/\mathbb{Z})$ to be the space of all infinitely differentiable periodic functions $f : \mathbb{R} \to \mathbb{C}$. For f and g in $C(\mathbb{R}/\mathbb{Z})$ let

$$\langle f, g\rangle \;=\; \int_0^1 f(x)\overline{g(x)}dx,$$

where the bar means complex conjugation, and the integral of a complex-valued function $h(x) = u(x) + iv(x)$ is defined by linearity, i.e.,

$$\int_0^1 h(x)dx = \int_0^1 u(x)dx + i\int_0^1 v(x)dx.$$

The reader who has up to now only seen integrals of functions from $\mathbb{R}$ to $\mathbb{R}$ should take a minute to verify that integrals of complex-valued functions satisfy the usual rules of calculus. These can be deduced from the real-valued case by splitting the function into real and imaginary part. For instance, if $f : [0, 1] \to \mathbb{C}$ is continuously differentiable, then $\int_0^1 f'(x)\, dx = f(1) - f(0)$.

Lemma 1.1.1 $\langle .,. \rangle$ *defines an inner product on the vector space* $C(\mathbb{R}/\mathbb{Z})$.

Proof: The linearity in the first argument is a simple exercise, and so is $\langle f,g \rangle = \overline{\langle g,f \rangle}$. For the positive definiteness recall that

$$\langle f,f \rangle = \int_0^1 |f(x)|^2 dx$$

is an integral over a real-valued and nonnegative function; hence it is real and nonnegative. For the last part let $f \neq 0$ and let $g(x) = |f(x)|^2$. Then g is a continuous function. Since $f \neq 0$, there is $x_0 \in [0,1]$ with $g(x_0) = \alpha > 0$. Then, since g is continuous, there is $\varepsilon > 0$ such that $g(x) > \alpha/2$ for every $x \in [0,1]$ with $|x - x_0| < \varepsilon$. This implies

$$\langle f,f \rangle = \int_0^1 g(x)dx \geq \int_{|x-x_0| < \varepsilon} \frac{\alpha}{2}dx \geq \varepsilon\alpha > 0.$$

$\square$

1.2 Exponentials

We shall now study the periodic exponential maps in more detail. For $k \in \mathbb{Z}$ let

$$e_k(x) = e^{2\pi i k x};$$

then e_k lies in $C(\mathbb{R}/\mathbb{Z})$. The inner products of the e_k are given in the following lemma.

Lemma 1.2.1 *If $k, l \in \mathbb{Z}$, then*

$$\langle e_k, e_l \rangle = \begin{cases} 1 & \text{if } k = l, \\ 0 & \text{if } k \neq l. \end{cases}$$

In particular, it follows that the e_k, for varying k, give linearly independent vectors in the vector space $C(\mathbb{R}/\mathbb{Z})$. Finally, if

$$f(x) = \sum_{k=-n}^{n} c_k e_k(x)$$

for some coefficients $c_k \in \mathbb{C}$, then

$$c_k = \langle f, e_k \rangle \quad \text{for each } k.$$

Proof: If $k = l$, then

$$\langle e_k, e_l \rangle = \int_0^1 e^{2\pi i k x} e^{-2\pi i k x} dx = \int_0^1 1 \, dx = 1.$$

Now let $k \neq l$ and set $m = k - l \neq 0$; then

$$\langle e_k, e_l \rangle = \int_0^1 e^{2\pi i m x} dx$$

$$= \frac{1}{2\pi i m} e^{2\pi i m x} \Big|_0^1$$

$$= \frac{1}{2\pi i m}(1 - 1) = 0.$$

From this we deduce the linear independence as follows. Suppose that we have

$$\lambda_{-n} e_{-n} + \lambda_{-n+1} e_{-n+1} + \cdots + \lambda_n e_n = 0$$

for some $n \in \mathbb{N}$ and coefficients $\lambda_k \in \mathbb{C}$. Then we have to show that all the coefficients λ_k vanish. To this end let k be an integer between $-n$ and n. Then

$$\begin{aligned}
0 &= \langle 0, e_k \rangle \\
&= \langle \lambda_{-n} e_{-n} + \cdots + \lambda_n e_n, e_k \rangle \\
&= \lambda_{-n} \langle e_{-n}, e_k \rangle + \cdots + \lambda_n \langle e_n, e_k \rangle \\
&= \lambda_k.
\end{aligned}$$

Thus the (e_k) are linearly independent, as claimed. In the same way we get $c_k = \langle f, e_k \rangle$ for f as in the theorem. $\square$

Let $f : \mathbb{R} \to \mathbb{C}$ be periodic and Riemann integrable on the interval $[0, 1]$. The numbers

$$c_k(f) = \langle f, e_k \rangle = \int_0^1 f(x) e^{-2\pi i k x} dx, \quad k \in \mathbb{Z},$$

are called the *Fourier coefficients* of f. The series

$$\sum_{k=-\infty}^{\infty} c_k(f) e^{2\pi i k x} = \sum_{k=-\infty}^{\infty} c_k(f) e_k(x),$$

i.e., the sequence of the partial sums

$$S_n(f) = \sum_{k=-n}^{n} c_k(f) e_k,$$

is called the *Fourier series* of f. Note that we have made no assertion on the convergence of the Fourier series so far. Indeed, it need not converge pointwise. We will show that it converges in the L^2-sense, a notion to be defined in the sequel.

Let $R(\mathbb{R}/\mathbb{Z})$ be the $\mathbb{C}$-vector space of all periodic functions $f : \mathbb{R} \to \mathbb{C}$ that are Riemann integrable on $[0,1]$. Since every continuous function on the interval $[0,1]$ is Riemann integrable, it follows that $C(\mathbb{R}/\mathbb{Z})$ is a subspace of $R(\mathbb{R}/\mathbb{Z})$. Note that the inner product $\langle .,. \rangle$ extends to $R(\mathbb{R}/\mathbb{Z})$, but it is no longer positive definite there (see Exercise 1.2).

For $f \in C(\mathbb{R}/\mathbb{Z})$ let

$$\|f\|_2 = \sqrt{\langle f, f \rangle}.$$

Then $\|.\|_2$ is a *norm* on the space $C(\mathbb{R}/\mathbb{Z})$; i.e.,

- it is multiplicative: $\|\lambda f\|_2 = |\lambda| \, \|f\|_2 \quad \lambda \in \mathbb{C}$,
- it is positive definite: $\|f\|_2 \geq 0$ and $\|f\|_2 = 0 \Rightarrow f = 0$,
- it satisfies the triangle inequality: $\|f + g\|_2 \leq \|f\|_2 + \|g\|_2$.

See Chapter 2 for a proof of this. Again the norm $\|.\|_2$ extends to $R(\mathbb{R}/\mathbb{Z})$ but loses its positive definiteness there.

1.3 The Bessel Inequality

The Bessel inequality gives an estimate of the sum of the square norms of the Fourier coefficients. It is of central importance in the theory of Fourier series. Its proof is based on the following lemma.

Lemma 1.3.1 *Let $f \in R(\mathbb{R}/\mathbb{Z})$, and for $k \in \mathbb{Z}$ let $c_k = \langle f, e_k \rangle$ be its kth Fourier coefficient. Then for all $n \in \mathbb{N}$,*

$$\left\| f - \sum_{k=-n}^{n} c_k e_k \right\|_2^2 = \|f\|_2^2 - \sum_{k=-n}^{n} |c_k|^2.$$

Proof: Let $g = \sum_{k=-n}^{n} c_k e_k$. Then

$$\langle f, g \rangle = \sum_{k=-n}^{n} \overline{c_k} \langle f, e_k \rangle = \sum_{k=-n}^{n} \overline{c_k} c_k = \sum_{k=-n}^{n} |c_k|^2,$$

and

$$\langle g, g \rangle = \sum_{k=-n}^{n} \overline{c_k} \langle g, e_k \rangle = \sum_{k=-n}^{n} |c_k|^2,$$

so that

$$\begin{aligned} \|f - g\|_2^2 &= \langle f - g, f - g \rangle \\ &= \langle f, f \rangle - \langle f, g \rangle - \langle g, f \rangle + \langle g, g \rangle \\ &= \|f\|_2^2 - \sum_{k=-n}^{n} |c_k|^2 - \sum_{k=-n}^{n} |c_k|^2 + \sum_{k=-n}^{n} |c_k|^2 \\ &= \|f\|_2^2 - \sum_{k=-n}^{n} |c_k|^2, \end{aligned}$$

which proves the lemma. $\qquad\square$

Theorem 1.3.2 *(Bessel inequality) Let $f \in R(\mathbb{R}/\mathbb{Z})$ with Fourier coefficients (c_k). Then*

$$\sum_{k=-\infty}^{\infty} |c_k|^2 \leq \int_0^1 |f(x)|^2 dx.$$

Proof: The lemma shows that for every $n \in \mathbb{N}$,

$$\sum_{k=-n}^{n} |c_k|^2 \leq \|f\|_2^2.$$

Let $n \to \infty$ to prove the theorem. $\qquad\square$

1.4 Convergence in the L^2-Norm

We shall now introduce the notion of L^2-convergence, which is the appropriate notion of convergence for Fourier series. Let f be in $R(\mathbb{R}/\mathbb{Z})$ and let f_n be a sequence in $R(\mathbb{R}/\mathbb{Z})$. We say that the sequence f_n *converges in the L^2-norm* to f if

$$\lim_{n\to\infty} \|f - f_n\|_2 = 0.$$

Note that if a sequence f_n converges to f in the L^2-norm, then it need not converge pointwise (see Exercise 1.4). Conversely, if a sequence converges pointwise, it need not converge in the L^2-norm (see Exercise 1.6).

A concept of convergence that indeed does imply L^2-convergence is that of *uniform convergence*. Recall that a sequence of functions f_n on an interval I converges uniformly to a function f if for every $\varepsilon > 0$ there is $n_0 \in \mathbb{N}$ such that for all $n \geq n_0$,

$$|f(x) - f_n(x)| < \varepsilon$$

for all $x \in I$. The difference between pointwise and uniform convergence lies in the fact that in the case of uniform convergence the number n_0 does not depend on x. It can be chosen uniformly for all $x \in I$.

Recall that if the sequence f_n converges uniformly to f, and all the functions f_n are continuous, then so is the function f.

Examples.

- The sequence $f_n(x) = x^n$ on the interval $I = [0,1]$ converges pointwise, but not uniformly, to the function

$$f(x) = \begin{cases} 0 & x < 1, \\ 1 & x = 1. \end{cases}$$

  However, on each subinterval $[0, a]$ for $a < 1$ the sequence converges uniformly to the zero function.

- Let $f_n(x) = \sum_{k=1}^{n} a_k(x)$ for a sequence of functions $a_k(x)$, $x \in I$. Suppose there is a sequence c_k of positive real numbers such that $|a_k(x)| \leq c_k$ for every $k \in \mathbb{N}$ and every $x \in I$. Suppose further that

$$\sum_{k \in \mathbb{N}} c_k < \infty.$$

  Then it follows that the sequence f_n converges uniformly to the function $f(x) = \sum_{k=1}^{\infty} a_k(x)$.

Proposition 1.4.1 *If the sequence f_n converges to f uniformly on $[0,1]$, then f_n converges to f in the L^2-norm.*

Proof: Let $\varepsilon > 0$. Then there is n_0 such that for all $n \geq n_0$,

$$|f(x) - f_n(x)| < \varepsilon \quad \text{for all } x \in [0,1].$$

Hence for $n \geq n_0$,

$$\|f - f_n\|_2^2 = \int_0^1 |f(x) - f_n(x)|^2 dx < \varepsilon^2,$$

so that $\|f - f_n\|_2 < \varepsilon$.   □

A key result of this chapter is that the Fourier series of every $f \in R(\mathbb{R}/\mathbb{Z})$ converges to f in the L^2-norm, which we shall now prove. The idea of the proof is to find a simple class of functions for which the claim can be proved by explicit calculation of the Fourier coefficients and to then approximate a given function by those simple ones. In order to carry out these explicit calculations we shall need the following lemma.

Lemma 1.4.2 *For* $0 \leq x \leq 1$ *we have*

$$\sum_{k=1}^{\infty} \frac{\cos 2\pi k x}{k^2} = \pi^2 \left(x^2 - x + \frac{1}{6} \right).$$

Note that as a special case for $x = 0$ we get Euler's formula

$$\sum_{k=1}^{\infty} \frac{1}{k^2} = \frac{\pi^2}{6}.$$

Proof: Let $\alpha < a < b < \beta$ be real numbers and let $f : [\alpha, \beta] \to \mathbb{R}$ be a continuously differentiable function. For $k \in \mathbb{R}$ let

$$F(k) = \int_a^b f(x) \sin(kx) dx.$$

Claim: $\lim_{|k| \to \infty} F(k) = 0$ and the convergence is uniform in $a, b \in [\alpha, \beta]$.

Proof of claim: For $t \neq 0$ we integrate by parts to get

$$F(k) = -f(x) \frac{\cos(kx)}{k} \Big|_a^b + \frac{1}{k} \int_a^b f'(x) \cos(kx) \, dx.$$

Since f and f' are continuous on $[\alpha, \beta]$, there is a constant $M > 0$ such that $|f(x)| \leq M$ and $|f'(x)| \leq M$ for all $x \in [\alpha, \beta]$. This implies

$$|F(k)| \leq \frac{2M}{|k|} + \frac{M(b-a)}{|k|},$$

which proves the claim.

We employ this as follows: Let $x \in (0,1)$. Since

$$2\pi \int_{\frac{1}{2}}^{x} \cos(2\pi kt)dt = \frac{\sin(2\pi kx)}{k}$$

and

$$\sum_{k=1}^{n} \cos(2\pi kx) = \frac{\sin((2n+1)\pi x)}{2\sin(\pi x)} - \frac{1}{2},$$

we get

$$\sum_{k=1}^{n} \frac{\sin(2\pi kx)}{k} = 2\pi \int_{\frac{1}{2}}^{x} \frac{\sin((2n+1)\pi t)}{2\sin(\pi t)}dt - \pi\left(x - \frac{1}{2}\right).$$

The first summand on the right-hand side tends to zero as $n \to \infty$ by the claim. This implies that for $0 < x < 1$,

$$\sum_{k=1}^{\infty} \frac{\sin(2\pi kx)}{k} = \pi\left(\frac{1}{2} - x\right),$$

and this series converges uniformly on the interval $[\delta, 1-\delta]$ for every $\delta > 0$. We now use this result to prove Lemma 1.4.2. Let

$$f(x) = \sum_{k=1}^{\infty} \frac{\cos(2\pi kx)}{k^2}.$$

We have just seen that the series of derivatives converges to $\pi^2(2x-1)$ and that this convergence is locally uniform, so for $0 < x < 1$ we have

$$f'(x) = \pi^2(2x - 1),$$

i.e., $f(x) = \pi^2(x^2 - x) + c$. We are left to show that $c = \frac{\pi^2}{6}$. Since the series defining f converges uniformly on $[0,1]$ and since $\int_0^1 \cos(2\pi kx)dx = 0$ for every $k \in \mathbb{N}$, we get

$$0 = \sum_{k=1}^{\infty} \int_0^1 \frac{\cos(2\pi kx)}{k^2}dx = \int_0^1 f(x)dx = \frac{\pi^2}{3} - \frac{\pi^2}{2} - c,$$

which implies that $c = \frac{\pi^2}{2} - \frac{\pi^2}{3} = \frac{\pi^2}{6}$.   □

Using this technical lemma we are now going to prove the convergence of the Fourier series for Riemannian step functions (see below) as follows.

For a subset A of $[0,1]$ let $\mathbf{1}_A$ be its *characteristic function*, i.e.,

$$\mathbf{1}_A(x) = \begin{cases} 1, & x \in A, \\ 0, & x \notin A. \end{cases}$$

Let $I_1, \ldots, I_m$ be subintervals of $[0,1]$ which can be open or closed or half-open. A *Riemann step function* is a function of the form

$$s(x) = \sum_{j=1}^{m} \alpha_j \mathbf{1}_{I_j}(x),$$

for some coefficients $\alpha_j \in \mathbb{R}$.

Recall the definition of the Riemann integral. First, for a Riemann step function $s(x) = \sum_{j=1}^m \alpha_j \mathbf{1}_{I_j}(x)$ one defines

$$\int_0^1 s(x)dx = \sum_{j=1}^m \alpha_j \text{ length}(I_j).$$

Recall that a real-valued function $f : [0,1] \to \mathbb{R}$ is called Riemann integrable if for every $\varepsilon > 0$ there are step functions φ and ψ on $[0,1]$ such that $\varphi(x) \le f(x) \le \psi(x)$ for every $x \in [0,1]$ and

$$\int_0^1 (\psi(x) - \varphi(x))\,dx < \varepsilon.$$

As ε shrinks to zero the integrals of the step functions will tend to a common limit, which is defined to be the integral of f. Note that as a consequence every Riemann integrable function on $[0,1]$ is bounded. A complex-valued function is called Riemann integrable if its real and imaginary parts are.

Lemma 1.4.3 *Let $f : \mathbb{R} \to \mathbb{R}$ be periodic and such that $f|_{[0,1]}$ is a Riemann step function. Then the Fourier series of f converges to f in the L^2-norm, i.e., the series*

$$f_n = S_n(f) = \sum_{k=-n}^n c_k e_k$$

converges to f in the L^2-norm, where for $k \in \mathbb{Z}$,

$$c_k = \int_0^1 f(x)e^{-2\pi ikx}dx.$$

Proof: By Lemma 1.3.1 it suffices to show that $\|f\|_2^2 = \sum_{k=-\infty}^\infty |c_k|^2$. First we consider the special case $f|_{[0,1]} = \mathbf{1}_{[0,a]}$ for some $a \in [0,1]$. The coefficients are $c_0 = a$, and

$$c_k = \int_0^a e^{-2\pi ikx}dx = \frac{i}{2\pi k}\left(e^{-2\pi ika} - 1\right)$$

for $k \neq 0$. Thus in the latter case we have

$$|c_k|^2 = \frac{1}{4\pi^2 k^2}(e^{2\pi ika} - 1)(e^{-2\pi ika} - 1) = \frac{1 - \cos(2\pi ka)}{2\pi^2 k^2}.$$

Using Lemma 1.4.2 we compute

$$
\begin{aligned}
\sum_{k=-\infty}^{\infty} |c_k|^2 &= a^2 + \sum_{k=1}^{\infty} \frac{1 - \cos(2\pi ka)}{\pi^2 k^2} \\
&= a^2 + \sum_{k=1}^{\infty} \frac{1}{\pi^2 k^2} - \frac{1}{\pi^2} \sum_{k=1}^{\infty} \frac{\cos(2\pi ka)}{k^2} \\
&= a^2 + \frac{1}{6} - \left(\frac{(1-2a)^2}{4} - \frac{1}{12} \right) \\
&= a \\
&= \int_0^1 |f(x)|^2 dx \\
&= \|f\|_2^2 .
\end{aligned}
$$

Therefore, we have proved the assertion of the lemma for the function $f = 1_{[0,a]}$. Next we shall deduce the same result for $f = 1_I$, where I is an arbitrary subinterval of $[0, 1]$. First note that neither the Fourier coefficients nor the norm changes if we replace the closed interval by the open or half-closed interval. Next observe the behavior of the Fourier coefficients under shifts; i.e., let $c_k(f)$ denote the kth Fourier coefficient of f and let $f^y(x) = f(x+y)$; then f^y is still periodic and Riemann integrable, and

$$
\begin{aligned}
c_k(f^y) &= \int_0^1 f^y(x) e^{-2\pi i k x} dx \\
&= \int_0^1 f(x+y) e^{-2\pi i k x} dx \\
&= \int_y^{1+y} f(x) e^{2\pi i k (y-x)} dx \\
&= e^{2\pi i k y} \int_0^1 f(x) e^{-2\pi i k x} dx \\
&= e^{2\pi i k y} c_k(f),
\end{aligned}
$$

since it doesn't matter whether one integrates a periodic function over $[0, 1]$ or over $[y, 1 + y]$. This implies $|c_k(f^y)|^2 = |c_k(f)|^2$. The same argument shows that $\|f^y\|_2 = \|f\|_2$, so that the lemma now follows for $f|_{[0,1]} = 1_I$ for an arbitrary interval in $[0, 1]$. An arbitrary step function is a linear combination of characteristic functions of intervals, so the lemma follows by linearity. □

Theorem 1.4.4 *Let $f : \mathbb{R} \to \mathbb{C}$ be periodic and Riemann integrable on $[0, 1]$. Then the Fourier series of f converges to f in the L^2-norm. If c_k denotes the Fourier coefficients of f, then*

$$\sum_{k=-\infty}^{\infty} |c_k|^2 = \int_0^1 |f(x)|^2 dx.$$

The theorem in particular implies that the sequence c_k tends to zero as $|k| \to \infty$. This assertion is also known as the *Riemann-Lebesgue Lemma*.

Proof: Let $f = u + iv$ be the decomposition of f into real and imaginary parts. The partial sums of the Fourier series for f satisfy $S_n(f) = S_n(u) + iS_n(v)$, so if the Fourier series of u and v converge in the L^2-norm to u and v, then the claim follows for f. To prove the theorem it thus suffices to consider the case where f is real-valued. Since, furthermore, integrable functions are bounded, we can multiply f by a positive scalar, so we may assume that $|f(x)| \le 1$ for all $x \in \mathbb{R}$.

Let $\varepsilon > 0$. Since f is Riemann integrable, there are step functions φ, ψ on $[0, 1]$ such that

$$-1 \le \varphi \le f \le \psi \le 1$$

and

$$\int_0^1 (\psi(x) - \varphi(x)) dx \le \frac{\varepsilon^2}{8}.$$

Let $g = f - \varphi$ then $g \ge 0$ and

$$|g|^2 \le |\psi - \varphi|^2 \le 2(\psi - \varphi),$$

so that

$$\int_0^1 |g(x)|^2 dx \le 2 \int_0^1 (\psi(x) - \varphi(x)) dx \le \frac{\varepsilon^2}{4}.$$

For the partial sums S_n we have

$$S_n(f) = S_n(\varphi) + S_n(g).$$

By Lemma 1.4.3 there is $n_0 \ge 0$ such that for $n \ge n_0$,

$$\|\varphi - S_n(\varphi)\|_2 \le \frac{\varepsilon}{2}.$$

By Lemma 1.3.1 we have the estimate

$$\|g - S_n(g)\|_2^2 \leq \|g\|_2^2 \leq \frac{\varepsilon^2}{4},$$

so that for $n \geq n_0$,

$$\|f - S_n(f)\|_2 \leq \|\varphi - S_n(\varphi)\|_2 + \|g - S_n(g)\|_2 \leq \frac{\varepsilon}{2} + \frac{\varepsilon}{2} = \varepsilon.$$

$\square$

1.5 Uniform Convergence of Fourier Series

Note that the last theorem does not tell us anything about pointwise convergence of the Fourier series. Indeed, the Fourier series does not necessarily converge pointwise to f. If, however, the function f is continuously differentiable, it does converge, as the next theorem shows, which is the second main result of this chapter.

Let $f : \mathbb{R} \to \mathbb{C}$ be continuous and periodic. We say that the function f is *piecewise continuously differentiable* if there are real numbers $0 = t_0 < t_1 < \cdots < t_r = 1$ such that for each j the function $f|_{[t_{j-1},t_j]}$ is continuously differentiable.

Theorem 1.5.1 *Let the function $f : \mathbb{R} \to \mathbb{C}$ be continuous, periodic, and piecewise continuously differentiable. Then the Fourier series of f converges uniformly to the function f.*

Proof: Let f be as in the statement of the theorem and let c_k denote the Fourier coefficients of f. Let $\varphi_j : [t_{j-1}, t_j] \to \mathbb{C}$ be the continuous derivative of f and let $\varphi : \mathbb{R} \to \mathbb{C}$ be the periodic function that for every j coincides with φ_j on the half-open interval $[t_{j-1}, t_j)$. Let γ_k be the Fourier coefficients of φ. Then

$$\sum_{k=-\infty}^{\infty} |\gamma_k|^2 \leq \|\varphi\|_2^2 < \infty.$$

Using integration by parts we compute

$$\int_{t_{j-1}}^{t_j} f(x)e^{-2\pi ikx} dx = \left. \frac{1}{-2\pi ik} f(x)e^{-2\pi ikx} \right|_{t_{j-1}}^{t_j}$$

$$-\frac{1}{-2\pi ik} \int_{t_{j-1}}^{t_j} \varphi(x)e^{-2\pi ikx} dx,$$

so that for $k \neq 0$ we obtain

$$c_k = \int_0^1 f(x)e^{-2\pi ikx}dx = \frac{1}{2\pi ik}\int_0^1 \varphi(x)e^{-2\pi ikx}dx = \frac{1}{2\pi ik}\gamma_k.$$

For $\alpha, \beta \in \mathbb{C}$ we have $0 \leq (|\alpha| - |\beta|)^2 = |\alpha|^2 + |\beta|^2 - 2|\alpha\beta|$ and thus $|\alpha\beta| \leq \frac{1}{2}(|\alpha|^2 + |\beta|^2)$, so that

$$|c_k| \leq \frac{1}{2}\left(\frac{1}{4\pi^2 k^2} + |\gamma_k|^2\right),$$

which implies

$$\sum_{k=-\infty}^{\infty} |c_k| < \infty.$$

Now, the final step of the proof is of importance in itself, and therefore we formulate it as a lemma.

Lemma 1.5.2 *Let f be continuous and periodic, and assume that the Fourier coefficients c_k of f satisfy*

$$\sum_{k=-\infty}^{\infty} |c_k| < \infty.$$

Then the Fourier series converges uniformly to f. In particular, we have for every $x \in \mathbb{R}$,

$$f(x) = \sum_{k \in \mathbb{Z}} c_k e_k(x).$$

Proof: The condition of the lemma implies that the Fourier series $\sum_{k=-\infty}^{\infty} c_k e^{2\pi ikx}$ converges uniformly. Denote the limit function by g. Then the function g, being the uniform limit of continuous functions, must be continuous. Since the Fourier series also converges to f in the L^2-norm, it follows that

$$\|f - g\|_2 = 0.$$

Since f and g are continuous, the positive definiteness of the norm implies $f = g$, which concludes the proof of the lemma and the theorem.   □

1.6   Periodic Functions Revisited

We have introduced the space $C(\mathbb{R}/\mathbb{Z})$ as the space of continuous periodic functions on $\mathbb{R}$. There is also a different interpretation of it, as follows. Firstly, on $\mathbb{R}$ establish the following equivalence relation:

$$x \sim y \;\Leftrightarrow\; x - y \in \mathbb{Z}.$$

For $x \in \mathbb{R}$ its equivalence class is $[x] = x + \mathbb{Z} = \{x + k | k \in \mathbb{Z}\}$. Let $\mathbb{R}/\mathbb{Z}$ be the set of all equivalence classes. This set can be identified with the half-open interval $[0, 1)$. It also can be identified with the unit torus

$$\mathbb{T} \;=\; \{z \in \mathbb{C} : |z| = 1\},$$

since the map $e : \mathbb{R} \to \mathbb{T}$ that maps x to $e(x) = e^{2\pi i x}$ gives a bijection between $\mathbb{R}/\mathbb{Z}$ and $\mathbb{T}$.

A sequence $[x_n]$ is said to *converge* to $[x] \in \mathbb{R}/\mathbb{Z}$ if there are representatives $x_n' \in \mathbb{R}$ and $x \in \mathbb{R}$ for the classes $[x_n]$ and $[x]$ such that the sequence (x_n') converges to x' in $\mathbb{R}$. In the interval $[0, 1)$ this means that either x_n converges to x in the interval $[0, 1)$ or that $[x] = 0$ and the sequence x_n decomposes into two subsequences one of which converges to 0 and the other to 1.

The best way to visualize $\mathbb{R}/\mathbb{Z}$ is as the real line "rolled up" by either identifying the integers or by using the map $e^{2\pi i x}$ or by gluing the ends of the interval $[0, 1]$ together.

Given the notion of convergence it is easy to say what a continuous function is. A function $f : \mathbb{R}/\mathbb{Z} \to \mathbb{C}$ is said to be *continuous* if for every convergent sequence $[x_n]$ in $\mathbb{R}/\mathbb{Z}$ the sequence $f([x_n])$ converges in $\mathbb{C}$.

Each continuous function on $\mathbb{R}/\mathbb{Z}$ can be composed with the natural projection $P : \mathbb{R} \to \mathbb{R}/\mathbb{Z}$ to give a continuous periodic function on $\mathbb{R}$. In this way we can identify $C(\mathbb{R}/\mathbb{Z})$ with the space of all continuous functions on $\mathbb{R}/\mathbb{Z}$, and we will view $C(\mathbb{R}/\mathbb{Z})$ in this way from now on.

1.7   Exercises

Exercise 1.1 Let $f : \mathbb{R} \to \mathbb{C}$ be continuous, periodic, and even, i.e., $f(-x) = f(x)$ for every $x \in \mathbb{R}$. Show that the Fourier series of f has the form

$$F_N(f) = c_0 + \sum_{k=1}^{N} 2c_k \cos(2\pi kx).$$

Exercise 1.2 Show by giving an example that the sesquilinear form $\langle ., . \rangle$ is not positive definite on the space $R(\mathbb{R}/\mathbb{Z})$.

Exercise 1.3 Let $f \in C(\mathbb{R}/\mathbb{Z})$. For $y > 0$ let

$$\omega(y) = \int_0^1 |f(t+y) - f(t)| dt.$$

Show that the Fourier coefficients c_k of f satisfy

$$|c_k| \leq \frac{1}{2}\omega\left(\frac{1}{2k}\right)$$

for $k \neq 0$.

(Hint: Use the fact that $c_k = - \int_0^1 f(t)e^{-2\pi ik(t-\frac{1}{2k})} dt$.)

Exercise 1.4 Show by example that a sequence f_n of integrable functions on $[0,1]$ that converges in the L^2-norm need not converge pointwise.

(Hint: Define f_n to be the characteristic function of an interval I_n. Choose these intervals so that their lengths tend to zero as n tends to infinity and so that any $x \in [0,1]$ is contained in infinitely many of the I_n.)

Exercise 1.5 For $n \in \mathbb{N}$ let f_n be the continuous function on the closed interval $[0,1]$ that satisfies $f_n(0) = 1$, $f_n(\frac{1}{n}) = 0$, $f_n(1) = 0$ and that is linear between these points. Show that f_n converges to the zero function pointwise but not uniformly on the open interval $(0,1)$.

Exercise 1.6 Show by example that there is a sequence of integrable functions on $[0,1]$ that converges pointwise but not in the L^2-norm.

(Hint: Modify the example of Exercise 1.5.)

Exercise 1.7 Compute the Fourier series of the periodic function f given by $f(x) = |x|$ for $-\frac{1}{2} \leq x \leq \frac{1}{2}$.

Exercise 1.8 Compute the Fourier series of the periodic function that on $[0,1)$ is given by $f(x) = x$.

Exercise 1.9 Compute the Fourier series of the function given by

$$f(x) = |\sin(2\pi x)|.$$

Exercise 1.10 A trigonometric polynomial is a function of the form

$$g(x) = \sum_{k=-N}^{N} c_k e^{2\pi i k x}$$

for some $N \in \mathbb{N}$ and some coefficients $c_k \in \mathbb{C}$.

(a) Show that every $f \in C(\mathbb{R}/\mathbb{Z})$ can be uniformly approximated by continuous functions that are piecewise linear.

(b) Conclude from part (a) that every $f \in C(\mathbb{R}/\mathbb{Z})$ can be uniformly approximated by trigonometric polynomials.

Exercise 1.11 Let $f \in C^\infty(\mathbb{R}/\mathbb{Z})$ and let c_k be its Fourier coefficients. Show that the sequence c_k is rapidly decreasing; i.e., for each $N \in \mathbb{N}$ there is $d_N > 0$ such that for $k \neq 0$,

$$|c_k| \leq \frac{d_N}{|k|^N}.$$

(Hint: Compute the Fourier coefficients of the derivatives of f.)

Exercise 1.12 Let c_k, $k \in \mathbb{Z}$ be a rapidly decreasing sequence as in Exercise 1.11. Show that there is a function $f \in C^\infty(\mathbb{R}/\mathbb{Z})$ such that the c_k are the Fourier coefficients of f.

Exercise 1.13 Let f, g be in $R(\mathbb{R}/\mathbb{Z})$ and define their convolution by

$$f * g(x) = \int_0^1 f(x - y)g(y)dy.$$

Show that for every $f \in R(\mathbb{R}/\mathbb{Z})$ and every $k \in \mathbb{Z}$ we have $e_k * f = c_k(f)e_k$. Deduce from this that with

$$D_n = \sum_{k=-n}^{n} e_k$$

we have $D_n * f = S_n(f)$.

Exercise 1.14 Let $f \in R(\mathbb{R}/\mathbb{Z})$ and let

$$\sigma_n f = \frac{1}{n+1}(S_0(f) + \cdots + S_n(f)).$$

For $s \in \mathbb{R}$ let

$$f_s^*(x, y) = \frac{1}{2}(f(x+y) + f(x-y) - 2s).$$

Show that

$$\sigma_n f(x) = \int_0^1 f_s^*(x, y) F_n(y) dy + s,$$

where

$$F_n = \frac{1}{n+1}(D_0 + \cdots + D_n).$$

Exercise 1.15 Let $f \in R(\mathbb{R}/\mathbb{Z})$ and assume that for each $x \in \mathbb{R}$ the limit

$$f(x+0) + f(x-0) = \lim_{y \to 0}(f(x+y) + f(x-y))$$

exists. Show that for every $x \in \mathbb{R}$,

$$\lim_{n \to \infty} \sigma_n f(x) = \frac{1}{2}(f(x+0) + f(x-0)).$$

(Hint: Use Exercise 1.14 with $s = \frac{1}{2}(f(x+0) + f(x-0))$. Show that

$$F_n(x) = \frac{1}{n+1}\left(\frac{\sin((n+1)\pi x)}{\sin \pi x}\right)^2$$

and that $\int_0^1 F_n(x) \, dx = 1$. Then show that F_n is small away from 0.)

Exercise 1.16 Let $f : \mathbb{R}^n \to \mathbb{C}$ be infinitely differentiable and suppose that $f(x + k) = f(x)$ for every $k = (k_1, \ldots, k_n) \in \mathbb{Z}^n$. Show that

$$f(x) = \sum_{k \in \mathbb{Z}^n} c_k e^{2\pi i \langle x, k \rangle},$$

where $\langle x, k \rangle = x_1 k_2 + \cdots + x_n k_n$ and

$$c_k = \int_0^1 \cdots \int_0^1 f(y) e^{-2\pi i \langle y, k \rangle} dy_1 \cdots dy_n.$$

Exercise 1.17 Let $k : \mathbb{R}^2 \to \mathbb{C}$ be smooth (i.e., infinitely differentiable) and invariant under the natural action of $\mathbb{Z}^2$; i.e., $k(x+k, y+l) = k(x, y)$ for all $k, l \in \mathbb{Z}$ and $x, y \in \mathbb{R}$. For $\varphi \in C(\mathbb{R}/\mathbb{Z})$ set

$$K\varphi(x) = \int_0^1 k(x, y)\varphi(y)dy.$$

Show that K satisfies

$$\|K\varphi\|_2^2 \leq \|\varphi\|_2^2 \int_0^1 \int_0^1 |k(x, y)|^2 dx \, dy.$$

Show that the sum

$$\operatorname{tr} K = \sum_{k \in \mathbb{Z}} \langle K e_k, e_k \rangle$$

converges absolutely and that

$$\operatorname{tr} K = \int_0^1 k(x, x) dx.$$

Chapter 2

Hilbert Spaces

In this chapter we shall reinterpret the results of the previous one in terms of Hilbert spaces, since this is the appropriate setting for the generalizations of the results of Fourier theory, that will be given in the chapters to follow.

2.1 Pre-Hilbert and Hilbert Spaces

A complex vector space V together with an inner product $\langle .,. \rangle$, is called a *pre-Hilbert space*. Other authors sometimes use the term *inner product space*, but since our emphasis is on Hilbert spaces, we shall use the term given.

Examples. The simplest example, besides the zero space, is $V = \mathbb{C}$ with $\langle \alpha, \beta \rangle = \alpha \bar{\beta}$.

A more general example is $V = \mathbb{C}^k$ for a natural number k with

$$\langle v, w \rangle = v^t \bar{w},$$

where we consider elements of $\mathbb{C}^k$ as column vectors, and where v^t is the transpose of v and $\bar{w}$ is the vector with complex conjugate entries. Using coordinates this means

$$\langle v, w \rangle = \left\langle \begin{pmatrix} v_1 \\ \vdots \\ v_k \end{pmatrix}, \begin{pmatrix} w_1 \\ \vdots \\ w_k \end{pmatrix} \right\rangle = v_1 \overline{w_1} + v_2 \overline{w_2} + \cdots + v_k \overline{w_k}.$$

It is a result of linear algebra that every finite-dimensional pre-Hilbert space V is isomorphic to $\mathbb{C}^k$ for $k = \dim V$.

Given a pre-Hilbert space V we define

$$\|v\| = \sqrt{\langle v, v \rangle}, \qquad \text{for } v \in V.$$

Lemma 2.1.1 *(Cauchy-Schwarz inequality) Let V be an arbitrary pre-Hilbert space. Then for every $v, w \in V$,*

$$|\langle v, w \rangle| \leq \|v\| \, \|w\|.$$

This implies that $\|.\|$ is a norm, i.e.,

- *it is multiplicative: $\|\lambda v\| = |\lambda| \, \|v\| \quad \lambda \in \mathbb{C}$,*

- *it is positive definite: $\|v\| \geq 0$; and $\|v\| = 0 \Rightarrow v = 0$,*

- *it satisfies the triangle inequality: $\|v + w\| \leq \|v\| + \|w\|$.*

Proof: Let $v, w \in V$. For every $t \in \mathbb{R}$ we define $\varphi(t)$ by

$$\varphi(t) = \|v\|^2 + t^2 \|w\|^2 + t(\langle v, w \rangle + \langle w, v \rangle).$$

We then have that

$$\varphi(t) = \langle v + tw, v + tw \rangle = \|v + tw\|^2 \geq 0.$$

Note that $\langle v, w \rangle + \langle w, v \rangle = 2\operatorname{Re} \langle v, w \rangle$. The real-valued function $\varphi(t)$ is a quadratic polynomial with positive leading coefficient. Therefore it takes its minimum value where its derivative φ' vanishes, i.e., at the point $t_0 = -\operatorname{Re} \langle v, w \rangle / \| w \|^2$. Evaluating at t_0, we see that

$$0 \leq \varphi(t_0) = \|v\|^2 + \frac{(\operatorname{Re} \langle v, w \rangle)^2}{\|w\|^2} - 2\frac{(\operatorname{Re} \langle v, w \rangle)^2}{\|w\|^2},$$

which implies $(\operatorname{Re} \langle v, w \rangle)^2 \leq \|v\|^2 \|w\|^2$. Replacing v by $e^{i\theta} v$ for a suitable real number θ establishes the initial claim.

We now show that this result implies the triangle inequality. We use the fact that for every complex number z we have $\operatorname{Re}(z) \leq |z|$, so

$$\begin{aligned}
\|v + w\|^2 &= \langle v + w, v + w \rangle \\
&= \|v\|^2 + \|w\|^2 + 2\operatorname{Re}(\langle v, w \rangle) \\
&\leq \|v\|^2 + \|w\|^2 + 2|\langle v, w \rangle| \\
&\leq \|v\|^2 + \|w\|^2 + 2\|v\| \, \|w\| \\
&= (\|v\| + \|w\|)^2.
\end{aligned}$$

Taking square roots of both sides gives the triangle inequality. The other conditions for $\|.\|$ to be a norm are obviously satisfied. $\qquad\square$

Lemma 2.1.2 *For every two* $v, w \in V$,

$$\big| \, \|v\| - \|w\| \, \big| \; \leq \; \|v - w\|.$$

Proof: The triangle inequality implies

$$\|v\| \; = \; \|v - w + w\| \; \leq \; \|v - w\| + \|w\|,$$

or

$$\|v\| - \|w\| \; \leq \; \|v - w\|.$$

Interchanging v and w gives

$$\|w\| - \|v\| \; \leq \; \|w - v\| \; = \; \|v - w\|.$$

Taken together, these two estimates prove the claim. $\qquad\qquad\square$

A linear map $T : V \to W$ between two pre-Hilbert spaces is called an *isometry* if T preserves inner products, i.e., if for all $v, v' \in V$,

$$\langle T(v), T(v') \rangle \; = \; \langle v, v' \rangle,$$

where the inner product on the left-hand side is the one on W, and on the right-hand side is the one on V. It follows that T must be injective, since if $T(v) = 0$, then

$$\langle v, v \rangle \; = \; \langle T(v), T(v) \rangle \; = \; \langle 0, 0 \rangle \; = \; 0,$$

which implies $v = 0$. Furthermore, if T is surjective, then T has a linear inverse $T^{-1} : W \to V$, which also is an isometry. In this case we say that T is a *unitary* map or an *isomorphism of pre-Hilbert spaces*.

Let $(V, \langle ., . \rangle)$ be a pre-Hilbert space. The property that makes a pre-Hilbert space into a Hilbert space is *completeness*. (Recall that it is completeness that distinguishes the real numbers from the rationals.) We will formulate the notion of completeness here in a similar fashion as in the passage from the rationals to the reals, i.e., as convergence of Cauchy sequences.

We say that a sequence $(v_n)_n$ in V *converges* to $v \in V$, if the sequence $\|v_n - v\|$ of real numbers tends to zero; in other words, if for every $\varepsilon > 0$ there is a natural number n_1 such that for every $n \geq n_1$ the estimate

$$\|v - v_n\| \; < \; \varepsilon$$

holds. In this case the vector v is uniquely determined by the sequence (v_n) and we write

$$v = \lim_{n \to \infty} v_n.$$

A subset D of a pre-Hilbert space H is called a *dense subset*, if every $h \in H$ is a limit of a sequence in D, i.e., if for any given $h \in H$ there is a sequence d_j in D with $\lim_{j \to \infty} d_j = h$. For example, the set $\mathbb{Q} + i\mathbb{Q}$ of all $a + bi$ with $a, b \in \mathbb{Q}$, is dense in $\mathbb{C}$.

A *Cauchy sequence* in V is a sequence $v_n \in V$ such that for every $\varepsilon > 0$ there is a natural number n_0 such that for every pair of natural numbers $n, m \geq n_0$, we have

$$\|v_n - v_m\| < \varepsilon.$$

It is easy to see that if (v_n), (w_n) are Cauchy sequences, then their sum $(v_n + w_n)$ is a Cauchy sequence. Further, if (v_n) converges to v and (w_n) converges to w, then $(v_n + w_n)$ converges to $v + w$ (see Exercise 2.5).

Lemma 2.1.3 *Every convergent sequence is Cauchy.*

Proof: Let (v_n) be a sequence in V convergent to $v \in V$. Let $\varepsilon > 0$ and let n_1 be a natural number such that for all $n \geq n_1$ we have

$$\|v - v_n\| < \frac{\varepsilon}{2}.$$

Let $n, m \geq n_1$. Then

$$
\begin{aligned}
\|v_n - v_m\| &= \|v_n - v + v - v_m\| \\
&\leq \|v_n - v\| + \|v_m - v\| < \frac{\varepsilon}{2} + \frac{\varepsilon}{2} = \varepsilon.
\end{aligned}
$$

$\square$

We call the space V a *complete* space or a *Hilbert space* if the converse of the above lemma is true, i.e., if every Cauchy sequence in V converges.

Actually, the notion of a Cauchy sequence and completeness only depends on the norm and does not directly relate to the inner product. A *normed space* is a complex vector space V together with a

norm $\|.\| : V \to [0, \infty)$, i.e. the map $\|.\|$ satisfies the three axioms in
Lemma 2.1.1. A normed space $(V, \|.\|)$ is called a *Banach space* if it
is complete, i.e., if every Cauchy sequence in V converges.

Proposition 2.1.4 *A pre-Hilbert space that is finite-dimensional, is
complete, i.e., is a Hilbert space.*

Proof: We prove this result by induction on the dimension. For a
zero-dimensional Hilbert space there is nothing to show. So let V be
a pre-Hilbert space of dimension $k+1$ and assume that the claim has
been proven for all spaces of dimension k. Let $v \in V$ be a nonzero
vector of norm 1. Let $W = \mathbb{C}v$ and let U be its *orthogonal space*,
i.e., the space of all $u \in V$ with $\langle u, v \rangle = 0$. Then V is the orthogonal
direct sum of W and U (see Exercise 2.10) and the dimension of U
is k, so this space is complete by the induction hypothesis.

Let (v_n) be a Cauchy sequence in V; then for each natural number
n,

$$v_n = \lambda_n v + u_n,$$

where λ_n is a complex number and $u_n \in U$. For $m, n \in \mathbb{N}$ we have

$$\|v_n - v_m\|^2 = |\lambda_n - \lambda_m|^2 + \|u_n - u_m\|^2,$$

so it follows that $|\lambda_n - \lambda_m| \leq \|v_n - v_m\|$ and since (v_n) is a Cauchy
sequence we derive that (λ_n) is a Cauchy sequence in $\mathbb{C}$, and thus is
convergent. Similarly we get that (u_n) is a Cauchy sequence in U,
which then also is convergent. Thus (v_n) is the sum of two convergent
sequences in V, and hence is also convergent. $\qquad\qquad\square$

2.2 ℓ^2-Spaces

We next introduce an important class of Hilbert spaces that gives
universal examples. These are called the ℓ^2-spaces. Let S be an
arbitrary set. Let $\ell^2(S)$ be the set of functions $f : S \to \mathbb{C}$ satisfying

$$\|f\|^2 = \sum_{s \in S} |f(s)|^2 < \infty.$$

The fact that the sum is finite actually means that all but countably
many of the $f(s)$ are zero, and that the sum over those countably

many converges absolutely. Another way to read the sum (see Exercise 2.6) is

$$\sum_{s \in S} |f(s)|^2 \;=\; \sup_{\substack{F \subset S \\ F \text{ finite}}} \sum_{s \in F} |f(s)|^2.$$

Note that if S is a finite set, then the convergence condition is vacuous, and so $\ell^2(S)$ then consists of the finite-dimensional complex vector space of all maps from S to $\mathbb{C}$. By Proposition 2.1.4 it therefore follows that $\ell^2(S)$ is a Hilbert space.

Theorem 2.2.1 *Let S be any set. Then $\ell^2(S)$ forms a Hilbert space with inner product*

$$\langle f, g \rangle \;=\; \sum_{s \in S} f(s)\overline{g(s)}, \qquad f, g \in \ell^2(S).$$

Proof: Let S be a set. First we must show that the inner product actually converges, i.e., we have to show that for every $f, g \in \ell^2(S)$ we have

$$\sum_{s \in S} |f(s)\overline{g(s)}| \;<\; \infty.$$

Once this has been established, the proof of the Cauchy-Schwarz inequality applies. From this one then infers the triangle inequality: $\|f + g\| \le \|f\| + \|g\|$, which means that $f, g \in \ell^2(S)$ implies $f + g \in \ell^2(S)$, so that $\ell^2(S)$ is a complex vector space. The fact that it is a pre-Hilbert space is then immediate.

Let us first prove the convergence of the scalar product. Since $|f(s)\overline{g(s)}| = |f(s)||g(s)|$, it suffices to prove the claim for real-valued nonnegative functions f and g. Let F be a finite subset of S. There are no convergence problems for $\ell^2(F)$; hence the latter is a Hilbert space and the Cauchy-Schwarz inequality holds for elements of $\ell^2(F)$. Let $f, g \in \ell^2(S)$ be real-valued and nonnegative and let f_F and g_F be their restrictions to F, which lie in $\ell^2(F)$. We have $\|f_F\| \le \|f\|$ and the same for g. We have the estimate

$$\sum_{s \in F} f(s)g(s) \;=\; |\langle f_F, g_F \rangle| \;\le\; \|f_F\|\,\|g_F\| \;\le\; \|f\|\,\|g\|\,.$$

This implies

$$\sum_{s \in S} f(s)g(s) \;=\; \sup_{\substack{F \subset S \\ F \text{ finite}}} \sum_{s \in F} f(s)g(s) \;\le\; \|f\|\,\|g\| \;<\; \infty.$$

So the convergence of the inner product is established, and by what was said above it follows that $\ell^2(S)$ is a pre-Hilbert space.

We are left to show that $\ell^2(S)$ is complete. For this let (f_n) be a Cauchy sequence in $\ell^2(S)$. Then for every $s_0 \in S$,

$$|f_n(s_0) - f_m(s_0)|^2 \leq \sum_{s \in S} |f_n(s) - f_m(s)|^2 = \|f_n - f_m\|^2,$$

which implies that $f_n(s_0)$ is a Cauchy sequence in $\mathbb{C}$, and hence is convergent to some complex number, $f(s_0)$ say. This means that the sequence of functions (f_n) converges pointwise to some function f on S.

Let $\varepsilon > 0$ and let $N \in \mathbb{N}$ be so large that for $m, n \geq N$ we have $\|f_n - f_m\|^2 < \varepsilon$. For $n \geq N$ and $F \subset S$ finite,

$$\sum_{s \in F} |f_n(s) - f(s)|^2 = \lim_{j \to \infty} \sum_{s \in F} |f_n(s) - f_j(s)|^2$$

$$\leq \sup_{j \geq N} \|f_n - f_j\|^2 \leq \varepsilon.$$

Therefore, for $n \geq N$,

$$\|f_n - f\|^2 = \sup_{F \subset S, \text{ finite}} \sum_{s \in F} |f_n(s) - f(s)|^2 \leq \varepsilon.$$

This implies that $f \in \ell^2(S)$ and that $f_n \to f$ in $\ell^2(S)$, so this space is complete. $\qquad\square$

It can actually be shown that every Hilbert space is isomorphic to one of the form $\ell^2(S)$ for some set S and that two spaces $\ell^2(S)$ and $\ell^2(S')$ are isomorphic if and only if S and S' have the same cardinality. However we will not go into this here, since we are interested only in separable Hilbert spaces, a notion to be introduced in the next section.

2.3   Orthonormal Bases and Completion

A *complete system* in a pre-Hilbert space H is a family $(a_j)_{j \in J}$ of vectors in H such that the linear subspace span(a_j) spanned by the a_j is dense in H. A pre-Hilbert space is called *separable* if it contains a countable complete system. (Here countable means either finite or countably infinite.)

Examples.

- For a finite dimensional Hilbert space any family that contains a basis is a complete system.

- To give an example of an infinite-dimensional separable Hilbert space consider the space $\ell^2(\mathbb{N})$. For $j \in \mathbb{N}$ let $\psi_j \in \ell^2(\mathbb{N})$ be defined by

$$\psi_j(k) = \begin{cases} 1 & \text{if } k = j, \\ 0 & \text{otherwise.} \end{cases}$$

  Then for every $f \in \ell^2(\mathbb{N})$ we get

$$\langle f, \psi_j \rangle = f(j),$$

  which implies that $(\psi_j)_{j\in\mathbb{N}}$ is indeed a complete system.

An *orthonormal system* in a pre-Hilbert space H is a family $(h_j)_{j\in J}$ of vectors in H such that for every $j, j' \in J$ we have $\langle h_j, h_{j'} \rangle = \delta_{j,j'}$, where $\delta_{j,j'}$ is the *Kronecker delta*:

$$\delta_{j,j'} = \begin{cases} 1, & j = j', \\ 0, & \text{otherwise.} \end{cases}$$

An orthonormal system is called *orthonormal basis*, if it is also a complete system.

Example. The system (ψ_j) above forms an orthonormal basis of the Hilbert space $\ell^2(\mathbb{N})$.

Proposition 2.3.1 *Every separable pre-Hilbert space H admits an orthonormal basis.*

The assertion also holds for nonseparable spaces, but the proof of that requires set-theoretic methods, and will not be given here.

Proof: The method used here is called Gram-Schmidt orthonormalization. For finite-dimensional spaces this is usually a feature of a linear algebra course.

Let $(a_j)_{j\in\mathbb{N}}$ be a complete system. If some a_j can be represented as a finite linear combination of the $a_{j'}$ with $j' < j$, then we can leave this element out and still keep a complete system. Thus we may assume that every finite set of the a_j is linearly independent. We then construct an orthonormal basis out of the a_j by an inductive procedure. First let

$$e_1 = \frac{a_1}{\|a_1\|}.$$

Next assume that $e_1, \ldots, e_k$ have already been constructed, being orthonormal and with $\mathrm{Span}(e_1, \ldots, e_k) = \mathrm{Span}(a_1, \ldots, a_k)$. Then put

$$e'_{k+1} = a_{k+1} - \sum_{j=1}^{k} \langle a_{k+1}, e_j \rangle\, e_j.$$

For $j = 1, \ldots, k$ then $\langle e'_{k+1}, e_j \rangle = 0$. Further, the linear independence implies that e'_{k+1} cannot be zero, so put

$$e_{k+1} = \frac{e'_{k+1}}{\|e'_{k+1}\|}.$$

Then $e_1, \ldots, e_{k+1}$ are orthonormal.

If H is finite-dimensional, this procedure will produce a basis (e_j) in finitely many steps and then stop. If H is infinite-dimensional, it will not stop and will thus produce a sequence $(e_j)_{j \in \mathbb{N}}$.

By construction we have $\mathrm{span}(e_j)_j = \mathrm{span}(a_j)_j$, which is dense in H. Therefore $(e_j)_{j \in \mathbb{N}}$ is an orthonormal basis. $\qquad\square$

Theorem 2.3.2 *Suppose H is an infinite-dimensional separable pre-Hilbert space; and let (e_j) be an orthonormal basis of H. Then every element h of H can be represented in the form*

$$h = \sum_{j=1}^{\infty} c_j e_j,$$

where the sum is convergent in H, and the coefficients c_j satisfy

$$\sum_{j=1}^{\infty} |c_j|^2 < \infty.$$

The coefficients are unique and are given by $c_j = c_j(h) = \langle h, e_j \rangle$. The map $h \mapsto (c_j)_{j \in \mathbb{N}}$ gives an isometry from H to $\ell^2(\mathbb{N})$. For $h, h' \in H$ we have

$$\langle h, h' \rangle = \sum_{j=0}^{\infty} c_j(h)\overline{c_j(h')},$$

so in particular, $\|h\|^2 = \sum_{j=1}^{\infty} |c_j|^2$.

Proof: Let $h \in H$, define $c_j(h) = \langle h, e_j \rangle$, and for $n \in \mathbb{N}$ let $s_n(h) = \sum_{j=1}^n c_j e_j \in H$. We now repeat the calculation of the proof of Lemma 1.3.1:

$$
\begin{aligned}
0 &\leq \|h - s_n(h)\|^2 \\
&= \left\langle h - \sum_{j=1}^n c_j e_j, h - \sum_{j=1}^n c_j e_j \right\rangle \\
&= \|h\|^2 - \sum_{j=1}^n |c_j|^2.
\end{aligned}
$$

This implies $\sum_{j=1}^n |c_j|^2 \leq \|h\|^2$ for every n and therefore $\sum_{j=1}^\infty |c_j|^2 < \infty$.

We therefore obtain a linear map $T : H \to l^2(\mathbb{N})$ mapping h to the sequence $(c_j(h))_j$. Since $\sum_{j=1}^n |c_j(h)|^2 \leq \|h\|^2$ we infer that $\|Th\| \leq \|h\|$ for every $h \in H$. For h in the span of $(e_j)_j$ we furthermore have $\|Th\| = \|h\|$. Since this subspace is dense, the latter equality holds for every $h \in H$ and so T is an isometry. In particular, $\langle h, h' \rangle = \langle Th, Th' \rangle = \sum_{j=1}^\infty c_j(h)\overline{c_j(h')}$. $\qquad \square$

The preceding theorem has several important consequences. Firstly, it shows that there is, up to isomorphism, only one separable Hilbert space of infinite dimension, namely $\ell^2(\mathbb{N})$. Secondly, it reduces all computations in a Hilbert space to computations with elements of an orthonormal basis. Finally, it allows us to embed a pre-Hilbert space as a dense subspace into a Hilbert space. To explain this further: A *dense subspace* of a pre-Hilbert space H is a subspace V such that for every $h \in H$ there is a sequence (v_n) in V converging to h.

Theorem 2.3.3 *(Completion) For every separable pre-Hilbert space V there is a Hilbert space H such that there is an isometry $T : V \to H$, called completion, that maps V onto a dense subspace of H. The completion is unique up to isomorphism in the following sense: If $T' : V \to H'$ is another isometry onto a dense subspace of a Hilbert space H', then there is a unique isomorphism of Hilbert spaces $S : H \to H'$ such that $T' = S \circ T$. We illustrate this by the following commutative diagram:*

It is customary to consider a pre-Hilbert space as a subspace of its completion by identifying it with the image of the completion map.

Again this theorem also holds for nonseparable spaces, but we prove it only for separable ones.

Proof: Let V be a separable pre-Hilbert space. If V is finite-dimensional, then V is itself a Hilbert space, and we can take T equal to the identity. Otherwise, choose an orthonormal basis (e_j), let $H = \ell^2(\mathbb{N})$, and let $T : V \to \ell^2(\mathbb{N})$ be the isometry given in Theorem 2.3.2. We have to show that $T(V)$ is dense in $H = \ell^2(\mathbb{N})$. Let $f \in \ell^2(\mathbb{N})$, and for $n \in \mathbb{N}$ let $f_n \in \ell^2(\mathbb{N})$ be given by

$$f_n(j) = \begin{cases} f(j) & \text{if } j \leq n, \\ 0 & \text{if } j > n. \end{cases}$$

Then

$$\|f - f_n\|^2 = \sum_{j>n} |f(j)|^2,$$

which tends to zero as n tends to infinity. So the sequence (f_n) converges to f in $\ell^2(\mathbb{N})$. For $j = 1, 2, \ldots, n$ let $\lambda_j = f(j)$. Then

$$f_n = T(\lambda_1 e_1 + \cdots + \lambda_n e_n),$$

so f_n lies in the image of T, which therefore is dense in H. This concludes the existence part of the proof.

For the uniqueness condition assume that there is a second isometry $T' : V \to H'$ onto a dense subspace. We define a map $S : H \to H'$ as follows: Let $h \in H$; then there is a sequence (v_n) in V such that $T(v_n)$ converges to h. Since T is an isometry it follows that (v_n) must be a Cauchy sequence in V, and since T' is an isometry the sequence $(T'(v_n))$ is Cauchy in H'. It therefore converges to some $h' \in H'$. We define $S(h) = h'$. It is easy to see that $S(h)$ does not depend on the choice of the sequence (v_n) and is therefore well-defined. To see that S is an isometry we let v, w be elements of H and choose sequences (v_n) and (w_n) in V such that $T(v_n)$ converges to v and $T(w_n)$ converges to w. We then compute

$$
\begin{aligned}
\langle S(v), S(w) \rangle &= \left\langle \lim_n T'(v_n), \lim_n T'(w_n) \right\rangle = \lim_n \langle v_n, w_n \rangle \\
&= \left\langle \lim_n T(v_n), \lim_n T(w_n) \right\rangle = \langle v, w \rangle .
\end{aligned}
$$

By construction S satisfies $T' = S \circ T$. □

Corollary 2.3.4 *Let V be a pre-Hilbert space with completion H, and let H' be a Hilbert subspace of H containing V. Then $H' = H$.*

Proof: Let $h \in H$. Then there is a sequence v_n in V converging to h. It follows that (v_n) must be a Cauchy sequence in $V \subset H'$, which is therefore convergent in H', and hence its limit h lies in H'. □

2.4   Fourier Series Revisited

In the previous chapter we saw that the space $C(\mathbb{R}/\mathbb{Z})$ with the inner product $\langle f, g \rangle = \int_0^1 f(x)\overline{g(x)}dx$ is a pre-Hilbert space. This space is not complete (see Exercise 2.12). Let

$$ L^2(\mathbb{R}/\mathbb{Z}) $$

denote its completion. Some of the main results of the first chapter can be summarized in the following theorem.

Theorem 2.4.1 *The exponentials $e_k(x) = e^{2\pi i k x}$, $k \in \mathbb{Z}$, form an orthonormal basis $(e_k)_{k\in\mathbb{Z}}$ of the Hilbert space $L^2(\mathbb{R}/\mathbb{Z})$.*

Proof: The orthonormality, i.e., $\langle e_k, e_{k'} \rangle = \delta_{k,k'}$, is given in Lemma 1.2.1. Hence the (e_k) form an orthonormal system. Let H be the space of all series of the form $\sum_{k\in\mathbb{Z}} c_k e_k$ with $\sum_{k\in\mathbb{Z}} |c_k|^2 < \infty$, which therefore converges in $L^2(\mathbb{R}/\mathbb{Z})$. Then the map $\sum_{k\in\mathbb{Z}} c_k e_k \mapsto (c_k)_{k\in\mathbb{Z}}$ gives an isomorphism to $\ell^2(\mathbb{Z})$, and hence H is a Hilbert subspace of $L^2(\mathbb{R}/\mathbb{Z})$. As a consequence of Theorem 1.4.4 it contains $C(\mathbb{R}/\mathbb{Z})$, and hence by Corollary 2.3.4 it equals $L^2(\mathbb{R}/\mathbb{Z})$. So we have shown that every element of $L^2(\mathbb{R}/\mathbb{Z})$ is representable as a sum $h = \sum_{k\in\mathbb{Z}} c_k e_k$. It follows that $c_k = \langle h, e_k \rangle$, and thus it emerges that (e_k) is complete. □

The space $L^2(\mathbb{R}/\mathbb{Z})$ can also be described as a space of (classes of) functions on $\mathbb{R}/\mathbb{Z}$ (see [20]). However, this requires techniques beyond the scope of this book, and it will not be pursued further.

2.5 Exercises

Exercise 2.1 Let H be a Hilbert space. Prove the following polarization identity for every $x, y \in H$:

$$4 \langle x, y \rangle = \|x + y\|^2 - \|x - y\|^2 + i \|x + iy\|^2 - i \|x - iy\|^2.$$

Exercise 2.2 Let $T: H \to H$ be a linear map which is continuous on a separable Hilbert space H. Show that the following are equivalent.

(a) T is unitary.

(b)   For every orthonormal basis (e_j) the family (Te_j) is an orthonormal basis again.

(c)   There is an orthonormal basis (e_j) such that the family (Te_j) is an orthonormal basis.

Exercise 2.3 Let H, H' be Hilbert spaces and let $T : H \to H'$ be a linear mapping such that $\|Tx\| = \|x\|$ for every $x \in H$. Show that T is an isometry, i.e., that for every $x, y \in H$:

$$\langle Tx, Ty \rangle = \langle x, y \rangle.$$

Exercise 2.4 Let $T: \ell^2(\mathbb{N}) \to \ell^2(\mathbb{N})$ be defined by

$$Tf(n) \overset{\text{def}}{=} \begin{cases} f(n-1) & n > 1, \\ 0 & n = 0. \end{cases}$$

Show that T is an isometry but not unitary.

Exercise 2.5 Show that if (v_n), (w_n) are Cauchy sequences, then their sum $(v_n + w_n)$ is a Cauchy sequence. Further, if (v_n) converges to v, and (w_n) converges to w, then $(v_n + w_n)$ converges to $v + w$.

Exercise 2.6 Let S be a set and f a nonnegative function on S. Suppose that $f(s)$ is zero except for s in a countable subset $\{s_1, s_2, \dots\} \subset S$. Show that

$$\sum_{j=1}^{\infty} f(s_j) = \sup_{\substack{F \subset S \\ F \text{ finite}}} \sum_{s \in F} f(s).$$

Note that both sides may be infinite.

Exercise 2.7 Let S be a set. Let $l^1(S)$ denote the set of all functions $f : S \to \mathbb{C}$ such that

$$\sum_{s \in S} |f(s)| < \infty.$$

Show that

$$\|f\|_1 = \sum_{s \in S} |f(s)|$$

defines a norm on $l^1(S)$.

Exercise 2.8 For which $s \in \mathbb{C}$ does the function $f(n) = n^{-s}$ belong to $\ell^2(\mathbb{N})$? For which does it belong to $l^1(\mathbb{N})$?

Exercise 2.9 For $T > 0$ let $C([-T,T])$ denote the space of all continuous functions $f : [-T,T] \to \mathbb{C}$. Show that the prescription

$$\langle f, g \rangle = \int_{-T}^{T} f(x)\overline{g(x)}dx$$

for $f, g \in C([-T,T])$ defines an inner product on this space.

Exercise 2.10 Let V be a finite-dimensional pre-Hilbert space and let $W \subset V$ be a subspace. Let U be the orthogonal space to W, i.e., U is the space of all $u \in V$ such that $\langle u, w \rangle = 0$ for every $w \in W$. Show that V is the direct sum of the subspaces W and U.

Exercise 2.11 Let H is a Hilbert space and let $(a_j)_{j \in J}$ be a family of elements of H. Show that (a_j) is a complete system if and only if its orthogonal space

$$(a_j)^\perp \overset{\text{def}}{=} \{h \in H : \langle h, a_j \rangle = 0 \text{ for every } j \in J\}$$

is the zero space.

Exercise 2.12 Show that the pre-Hilbert space $C(\mathbb{R}/\mathbb{Z})$ is not complete.

(Hint: For $n \in \mathbb{N}$ construct a function $f_n \in C(\mathbb{R}/\mathbb{Z})$ that takes values in $[0,1]$ and satisfies $f_n \equiv 0$ on $\left[0, \frac{1}{2} - \frac{1}{n+1}\right)$ as well as $f_n \equiv 1$ on $\left[\frac{1}{2}, 1\right)$.)

Exercise 2.13 Let E be a pre-Hilbert space and let (v_n) be a sequence in E. Show that (v_n) can converge to at most one element in E; i.e., show that if (v_n) converges to v and to v', then $v = v'$.

Exercise 2.14 Let $(V, \langle ., . \rangle)$ be a pre-Hilbert space. Show that the inner product is continuous; i.e., show that if the sequence (v_n) converges to v in V and the sequence (w_n) converges to w in V, then the sequence $\langle v_n, w_n \rangle$ converges to $\langle v, w \rangle$.

Exercise 2.15 Let (v_n) be a Cauchy sequence in some pre-Hilbert space V. Show that the sequence of norms $(\|v_n\|)$ forms a Cauchy sequence in $\mathbb{C}$.

Exercise 2.16 Let H be a Hilbert space and $v, w \in H$. Show that

$$\|v + w\|^2 + \|v - w\|^2 = 2\|v\|^2 + 2\|w\|^2.$$

This equality is known as the parallelogram law.

Exercise 2.17 Let H be a Hilbert space and let $T : H \to H$ be a map. Assume that T has an *adjoint*, i.e., there is a map T^* on H such that

$$\langle Tv, w \rangle = \langle v, T^* w \rangle$$

for all $v, w \in H$. Show that T and T^* are both linear.

Exercise 2.18 Let V be a finite-dimensional Hilbert space. A linear operator $A : V \to V$ is called self-adjoint if for any two vectors $v, w \in V$ we have

$$\langle Av, w \rangle = \langle v, Aw \rangle.$$

Show that if A is self-adjoint, then A is diagonalizable, i.e., that V has a basis consisting of eigenvectors of A.

(Hint: Show that if A leaves stable a subspace W of V, then it also leaves stable its orthogonal space $W^\perp$. Next make an induction on the dimension of V.)

Exercise 2.19 Let $C([0, 1])$ be the pre-Hilbert space of all continuous functions on the interval $[0, 1]$ with the inner product

$$\langle f, g \rangle = \int_0^1 f(x)\overline{g(x)}dx.$$

Let V be the subspace of all functions vanishing identically on $\left[0, \frac{1}{2}\right]$. Show that $V^\perp$ is the space of all functions vanishing on $\left[\frac{1}{2}, 1\right]$.

Exercise 2.20 Let F be a pre-Hilbert space, and let E be a dense subspace of F. Show that their completions coincide; i.e., show that every completion of F is a completion of E and that F can be embedded into every completion of E to make it a completion of F as well.

Chapter 3

The Fourier Transform

In the chapter on Fourier series we showed that every continuous periodic function can be written as a sum of simple waves. A similar result holds for nonperiodic functions on $\mathbb{R}$, provided that they are square integrable. In the periodic case the possible waves were $\cos(2\pi kx)$ and $\sin(2\pi kx)$ where k has to be an an integer, which means that the possible "wave lengths" are $1, \frac{1}{2}, \frac{1}{3}, \ldots$. In the nonperiodic case there is no restriction on the wavelengths, so every positive real number can occur. Consequently, the sum in the case of Fourier series will have to be replaced by an integral over $\mathbb{R}$, thus giving the Fourier transform.

3.1 Convergence Theorems

Before we arrive at the Fourier transform on $\mathbb{R}$ we will need two invaluable technical tools: the dominated convergence theorem and the monotone convergence theorem. We will here give only rather weak versions of these results. The interested reader is referred to [20] for more information on the subject.

Recall that a sequence of continuous functions f_n on $\mathbb{R}$ is said to converge *locally uniformly* to a function f if for every point $x \in \mathbb{R}$ there is a neighborhood on which f_n converges uniformly. This is equivalent to saying that the sequence converges uniformly on every closed interval $[a, b]$ for $a, b \in \mathbb{R}$ (see Exercise 3.10).

Lemma 3.1.1 *(This is a special case of the dominated convergence theorem.) Let f_n be a sequence of continuous functions on $\mathbb{R}$ that converges locally uniformly to some function f. Suppose that there is a nonnegative function g on $\mathbb{R}$ satisfying $\int_{-\infty}^{\infty} g(x)dx < \infty$ and $|f_n(x)| \leq g(x)$ for every $x \in \mathbb{R}$ and every $n \in \mathbb{N}$. Then the integrals $\int_{-\infty}^{\infty} f_n(x)dx$ and $\int_{-\infty}^{\infty} f(x)dx$ exist and*

$$\lim_{n\to\infty} \int_{-\infty}^{\infty} f_n(x)dx = \int_{-\infty}^{\infty} f(x)dx.$$

Proof: For every $T > 0$ the sequence f_n converges uniformly on $[-T, T]$. Therefore,

$$\int_{-T}^{T} |f(x)|dx = \lim_{n\to\infty} \int_{-T}^{T} |f_n(x)|dx \leq \int_{-T}^{T} g(x)dx$$
$$\leq \int_{-\infty}^{\infty} g(x)dx < \infty,$$

and for each n,

$$\int_{-T}^{T} |f_n(x)|dx \leq \int_{-T}^{T} g(x)dx \leq \int_{-\infty}^{\infty} g(x)dx < \infty,$$

which implies that the integrals exist. Let $g_n = f_n - f$. We have $|g_n| \leq 2g$, and we have to show that $\int_{-\infty}^{\infty} g_n(x)dx$ tends to zero. For this let $\varepsilon > 0$. Then there is $T > 0$ such that

$$\int_{|x|>T} 2g(x)dx < \frac{\varepsilon}{2}.$$

Next, since g_n tends to zero uniformly on $[-T, T]$, there is $n_0 \in \mathbb{N}$ such that for all $n \geq n_0$ we have

$$\int_{-T}^{T} |g_n(x)|dx < \frac{\varepsilon}{2}.$$

For $n \geq n_0$ we get

$$\int_{-\infty}^{\infty} |g_n(x)|dx = \int_{-T}^{T} |g_n(x)|dx + \int_{|x|>T} |g_n(x)|dx$$
$$\leq \int_{-T}^{T} |g_n(x)|dx + 2\int_{|x|>T} g(x)dx$$
$$< \frac{\varepsilon}{2} + \frac{\varepsilon}{2} = \varepsilon.$$

This concludes the proof of the lemma. $\square$

Lemma 3.1.2 *(This is a special case of the monotone convergence theorem.) Let f_n be a sequence of continuous nonnegative functions on $\mathbb{R}$ and assume that there is a continuous function f such that $f_n \to f$ locally uniformly and monotonically from below, i.e., $f_{n+1}(x) \geq f_n(x)$ for every $n \in \mathbb{N}$ and $x \in \mathbb{R}$. Then*

$$\lim_{n \to \infty} \int_{-\infty}^{\infty} f_n(x)dx = \int_{-\infty}^{\infty} f(x)dx.$$

Proof: If $\int_{-\infty}^{\infty} f(x)dx < \infty$, then the claim follows from the dominated convergence theorem, so let us assume that $\int_{-\infty}^{\infty} f(x)dx = \infty$. For every $C > 0$ there is $T > 0$ such that

$$\int_{-T}^{T} f(x)dx > C.$$

By locally uniform convergence there then is $n_0 \in \mathbb{N}$ such that for $n \geq n_0$,

$$\int_{-\infty}^{\infty} f_n(x)dx \geq \int_{-T}^{T} f_n(x)dx > C,$$

which implies the claim. $\qquad\square$

3.2 Convolution

Convolution is a standard technique that can be used, for example, to find smooth approximations of continuous functions. For us it is an essential tool in the proof of the main theorems of this chapter. Let $L_{\mathrm{bc}}^1(\mathbb{R})$ be the set of all bounded continuous functions $f : \mathbb{R} \to \mathbb{C}$ satisfying

$$\|f\|_1 \overset{\text{def}}{=} \int_{-\infty}^{\infty} |f(x)|\,dx < \infty.$$

It is easy to see that $\|.\|_1$ satisfies the axioms of a norm, i.e., that for $f, g \in L_{\mathrm{bc}}^1(\mathbb{R})$ and $\lambda \in \mathbb{C}$ we have

- $\|\lambda f\|_1 = |\lambda|\ \|f\|_1$,

- $\|f\|_1 = 0 \iff f = 0$, and

- $\|f + g\|_1 \leq \|f\|_1 + \|g\|_1$.

The last item, the triangle inequality, ensures that if f and g are in $L^1_{bc}(\mathbb{R})$, then so is their sum $f + g$, so $L^1_{bc}(\mathbb{R})$ is actually a vector space.

Theorem 3.2.1 *Let $f, g \in L^1_{bc}(\mathbb{R})$. Then the integral*

$$f * g(x) = \int_{-\infty}^{\infty} f(y)g(x - y)dy$$

exists for every $x \in \mathbb{R}$ and defines a function $f * g \in L^1_{bc}(\mathbb{R})$. The following equations hold for $f, g, h \in L^1_{bc}(\mathbb{R})$:

$$f*g = g*f, \quad f*(g*h) = (f*g)*h, \quad \text{and} \quad f*(g+h) = f*g+f*h.$$

The function $f * g$ is called the convolution, or convolution product, of the functions f and g.

Proof: Assume $|g(x)| \leq C$ for every $x \in \mathbb{R}$. Then

$$\int_{-\infty}^{\infty} |f(y)g(x - y)|dy \leq C \int_{-\infty}^{\infty} |f(y)|dy = C \|f\|_1,$$

which implies the existence and boundedness of $f * g$. Next we shall prove that it is continuous. Let $x_0 \in \mathbb{R}$. Assume $|f(x)|, |g(x)| \leq C$ for all $x \in \mathbb{R}$ and assume $g \neq 0$. For a given $\varepsilon > 0$ there is $T > |x_0|$ such that

$$\int_{|y|>T} |f(y)|dy < \frac{\varepsilon}{4C}.$$

Since a continuous function on a bounded closed interval is uniformly continuous, there is $\delta > 0$ such that

$$|x| \leq 2T, \quad |x - x'| < \delta \quad \Rightarrow \quad |g(x) - g(x')| < \frac{\varepsilon}{2\|g\|_1}.$$

Then for $|x - x_0| < \delta$ we have

$$\left| \int_{-T}^{T} f(y)g(x - y)dy - \int_{-T}^{T} f(y)g(x_0 - y)dy \right|$$

$$\leq \int_{-T}^{T} |f(y)||g(x - y) - g(x_0 - y)| \, dy$$

$$\leq \frac{\varepsilon}{2\|f\|_1} \int_{-T}^{T} |f(y)|dy \leq \frac{\varepsilon}{2}$$

and

$$\int_{|y|>T} |f(y)||g(x-y) - g(x_0 - y)|\, dy \ \leq\ 2C \int_{|y|>T} |f(y)|dy \ <\ \frac{\varepsilon}{2}.$$

Together these results imply that for $|x - x_0| < \delta$ we have

$$|f * g(x) - f * g(x_0)| \ <\ \varepsilon,$$

so $f * g$ is continuous at x_0.

To see that $\|f * g\|_1 < \infty$ we compute

$$
\begin{aligned}
\|f * g\|_1 \ &=\ \int_{-\infty}^{\infty} |f * g(x)|dx \\
&=\ \int_{-\infty}^{\infty} \left| \int_{-\infty}^{\infty} f(y)g(x-y)dy \right| dx \\
&\leq\ \int_{-\infty}^{\infty} \int_{-\infty}^{\infty} |f(y)g(x-y)|dy\, dx \\
&=\ \int_{-\infty}^{\infty} \int_{-\infty}^{\infty} |f(y)g(x-y)|dx\, dy \\
&=\ \int_{-\infty}^{\infty} |f(y)|dy \int_{-\infty}^{\infty} |g(x)|dx \ =\ \|f\|_1 \|g\|_1 .
\end{aligned}
$$

Next we show that $f * g = g * f$. The substitution $y \mapsto x - y$ gives

$$f * g(x) \ =\ \int_{-\infty}^{\infty} f(y)g(x-y)dy \ =\ \int_{-\infty}^{\infty} f(x-y)g(y)dy \ =\ g * f(x).$$

Further, since all integrals converge absolutely, we are allowed to change the order of integration in

$$
\begin{aligned}
f * (g * h)(x) \ &=\ \int_{-\infty}^{\infty} f(y) \int_{-\infty}^{\infty} g(z)h(x - y - z)\, dz\, dy \\
&=\ \int_{-\infty}^{\infty} g(z) \int_{-\infty}^{\infty} f(y)h(x - y - z)\, dy\, dz \\
&=\ \int_{-\infty}^{\infty} \int_{-\infty}^{\infty} f(y)g(z - y)h(x - z)\, dy\, dz \\
&=\ (f * g) * h(x).
\end{aligned}
$$

The distributive law $f * (g + h) = f * g + f * h$ is immediate. $\square$

3.3   The Transform

For $f \in L^1_{bc}(\mathbb{R})$ define its *Fourier transform* by

$$\hat{f}(y) = \int_{-\infty}^{\infty} f(x)e^{-2\pi ixy}dx.$$

By the estimate

$$|\hat{f}(y)| \leq \int_{-\infty}^{\infty} |f(x)e^{-2\pi ixy}|dx = \int_{-\infty}^{\infty} |f(x)|dx < \infty$$

it emerges that the Fourier transform $\hat{f}$ is bounded for every $f \in L^1_{bc}(\mathbb{R})$. We will now derive the first properties of the Fourier transform.

Theorem 3.3.1 *Let $f \in L^1_{bc}(\mathbb{R})$.*

(a) *If $g(x) = f(x)e^{2\pi iax}$ for $a \in \mathbb{R}$, then $\hat{g}(y) = \hat{f}(y - a)$.*

(b) *If $g(x) = f(x - a)$, then $\hat{g}(y) = \hat{f}(y)e^{-2\pi iay}$.*

(c) *If $g \in L^1_{bc}(\mathbb{R})$ and $h = f * g$, then $\hat{h}(y) = \hat{f}(y)\hat{g}(y)$.*

(d) *If $g(x) = f(\frac{x}{\lambda})$ for $\lambda > 0$, then $\hat{g}(y) = \lambda \hat{f}(\lambda y)$.*

(e) *If $g(x) = -2\pi ixf(x)$ and $g \in L^1_{bc}(\mathbb{R})$, then $\hat{f}$ is continuously differentiable with $\hat{f}'(y) = \hat{g}(y)$.*

(f) *Let f be continuously differentiable and assume that the functions f and f' lie in $L^1_{bc}(\mathbb{R})$. Then $\widehat{f'}(y) = 2\pi iy\hat{f}(y)$, so in particular, the function $y\hat{f}(y)$ is bounded.*

(g) *Let f be two times continuously differentiable and assume that the functions f, f', f'' are all in $L^1_{bc}(\mathbb{R})$. Then $\hat{f} \in L^1_{bc}(\mathbb{R})$.*

Proof: The points (a), (b) and (d) are direct consequences of the definition. For (c) we compute

$$
\begin{aligned}
\hat{h}(y) &= \int_{-\infty}^{\infty} h(x)e^{-2\pi i x y}\,dx \\
&= \int_{-\infty}^{\infty}\int_{-\infty}^{\infty} f(x-z)g(z)dz\, e^{-2\pi i x y}\,dx \\
&= \int_{-\infty}^{\infty}\int_{-\infty}^{\infty} f(x-z)e^{-2\pi i x y}\,dx\, g(z)dz \\
&= \int_{-\infty}^{\infty}\int_{-\infty}^{\infty} f(x)e^{-2\pi i x y}\,dx\, g(z)e^{-2\pi i z y}\,dz \\
&= \hat{f}(y)\hat{g}(y).
\end{aligned}
$$

For (e) note that

$$
\frac{\hat{f}(y)-\hat{f}(z)}{y-z} = \int_{-\infty}^{\infty} f(x)e^{-2\pi i z x}\frac{e^{-2\pi i x(y-z)}-1}{y-z}\,dx.
$$

Let $\varphi(x,u) = \frac{(e^{-2\pi i x u}-1)}{u}$. Then $|\varphi(x,u)| \le 2\pi|x|$ for all $u \neq 0$ and

$$
\varphi(x,u) \to -2\pi i x \quad \text{as } u \to 0,
$$

and the convergence is locally uniform in x. By dominated convergence the claim follows.

For (f) recall that by the integrability of $|f|$ there exist sequences $S_n, T_n \to \infty$ such that $f(-S_n), f(T_n) \to 0$. Then we compute

$$
\begin{aligned}
\widehat{f'}(y) &= \lim_{n\to\infty}\int_{-S_n}^{T_n} f'(x)e^{-2\pi i x y}\,dx \\
&= \lim_{n\to\infty}\left[f(T_n)e^{-2\pi i y T_n} - f(-S_n)e^{2\pi i y S_n}\right] \\
&\quad + \lim_{n\to\infty}\left[2\pi i y \int_{-S_n}^{T_n} f(x)e^{-2\pi i x y}\,dy\right] \\
&= 2\pi i y \int_{-\infty}^{\infty} f(x)e^{-2\pi i x y}\,dx.
\end{aligned}
$$

Finally, for (g) apply (f) twice to get that $y^2\hat{f}(y)$ is bounded. Since $\hat{f}$ is continuous, it is therefore integrable. $\qquad\square$

Lemma 3.3.2 *(Riemann-Lebesgue Lemma)*
Let $f \in L^1_{bc}(\mathbb{R})$. Then $\lim_{|x|\to\infty}\hat{f}(x) = 0$.

Proof: We compute

$$f(x) = \int_{-\infty}^{\infty} f(y) e^{-2\pi i x y} \, dy$$

$$= -\int_{-\infty}^{\infty} f(y) e^{-2\pi i x \left(y + \frac{1}{2x}\right)} \, dy$$

$$= -\int_{-\infty}^{\infty} f\left(y - \frac{1}{2x}\right) e^{-2\pi i x y} \, dy.$$

So we get

$$\hat{f}(x) = \frac{1}{2} \int_{-\infty}^{\infty} \left(f(y) - f\left(y - \frac{1}{2x}\right) \right) e^{-2\pi i x y} \, dy.$$

By dominated convergence and the continuity of f it follows that $\lim_{|x| \to \infty} \hat{f}(x) = 0$. $\qquad\square$

Let $S = S(\mathbb{R})$ be the space of *Schwartz functions*; i.e., S consists of all infinitely differentiable functions $f : \mathbb{R} \to \mathbb{C}$ such that for every $m, n \geq 0$ we have

$$\sigma_{m,n}(f) = \sup_{x \in \mathbb{R}} |x^m f^{(n)}(x)| < \infty.$$

An example for a Schwartz function is given by $f(x) = e^{-x^2}$.

Proposition 3.3.3 *We have $S \subset L^1_{bc}(\mathbb{R})$, and the Fourier transform maps S to itself; i.e., if $f \in S$, then $\hat{f} \in S$.*

Proof: Let $f \in S$. Then f is bounded and continuous and $(1 + x^2) f(x)$ is bounded, say by $C > 0$, so

$$\int_{-\infty}^{\infty} |f(x)| dx \leq C \int_{-\infty}^{\infty} \frac{1}{1 + x^2} dx < \infty,$$

and therefore f indeed lies in $L^1_{bc}(\mathbb{R})$. An iteration of Theorem 3.3.1 (e) gives that for every $f \in S$ we have that $\hat{f}$ is infinitely differentiable and that

$$((-2\pi i x)^n f)\hat{} = \hat{f}^{(n)}$$

for every $n \in \mathbb{N}$. Next, an iteration of Theorem 3.3.1 (f) shows that for every $f \in S$,

$$\widehat{f^{(n)}}(y) = (2\pi i y)^n \hat{f}(y)$$

for every $n \in \mathbb{N}$. Taking these together, we see that for every $f \in \mathcal{S}$ and every $m, n \geq 0$ the function

$$y^m \hat{f}^{(n)}(y)$$

is a Fourier transform of a function in $\mathcal{S}$, and hence it is bounded.

$\square$

3.4 The Inversion Formula

In this section we will show that the Fourier transform is, up to a sign twist, inverse to itself. We will need an auxiliary function as follows: For $\lambda > 0$ and $x \in \mathbb{R}$ let

$$h_\lambda(x) = \int_{-\infty}^{\infty} e^{-\lambda|t|} e^{2\pi i t x} dt.$$

Note that $0 < e^{-\lambda|t|} \leq 1$ and that $e^{-\lambda|t|}$ converges to 1 locally uniformly as $\lambda \to 0$.

Lemma 3.4.1 *We have*

$$h_\lambda(x) = \frac{2\lambda}{4\pi^2 x^2 + \lambda^2} \quad and \quad \int_{-\infty}^{\infty} h_\lambda(x) dx = 1.$$

In particular, it follows that $h_\lambda(x) = \frac{1}{\lambda} h_1(\frac{x}{\lambda})$ for every $\lambda > 0$.

Proof: We write

$$
\begin{aligned}
h_\lambda(x) &= \int_0^{\infty} e^{2\pi i t x - \lambda t} dt + \int_{-\infty}^0 e^{2\pi i t x + \lambda t} dt \\
&= \frac{e^{2\pi i t x - \lambda t}}{2\pi i x - \lambda} \Big|_0^{\infty} + \frac{e^{2\pi i t x + \lambda t}}{2\pi i x + \lambda} \Big|_{-\infty}^0 \\
&= \frac{1}{\lambda - 2\pi i x} + \frac{1}{\lambda + 2\pi i x} = \frac{2\lambda}{\lambda^2 + 4\pi^2 x^2}.
\end{aligned}
$$

Using Exercise 3.1 we get

$$
\begin{aligned}
\int_{-\infty}^{\infty} h_\lambda(x) dx &= \frac{2}{\lambda} \int_{-\infty}^{\infty} \frac{1}{1 + \left(\frac{2\pi x}{\lambda}\right)^2} dx \\
&= \frac{1}{\pi} \int_{-\infty}^{\infty} \frac{1}{1 + x^2} dx = 1. \quad \text{Q.E.D.}
\end{aligned}
$$

Lemma 3.4.2 *If $f \in L^1_{bc}(\mathbb{R})$, then for every $\lambda > 0$,*

$$f * h_\lambda(x) = \int_{-\infty}^{\infty} e^{-\lambda|t|} \hat{f}(t) e^{2\pi i x t} dt.$$

Proof: We compute

$$
\begin{aligned}
f * h_\lambda(x) &= \int_{-\infty}^{\infty} f(y) h_\lambda(x - y) \, dy \\
&= \int_{-\infty}^{\infty} f(y) \int_{-\infty}^{\infty} e^{-\lambda|t|} e^{2\pi i t(x-y)} \, dt \, dy \\
&= \int_{-\infty}^{\infty} e^{-\lambda|t|} e^{2\pi i x t} \int_{-\infty}^{\infty} f(y) e^{-2\pi i t y} \, dy \, dt.
\end{aligned}
$$

The claim follows.   □

Lemma 3.4.3 *For every $f \in L^1_{bc}(\mathbb{R})$ and every $x \in \mathbb{R}$ we have*

$$\lim_{\lambda \to 0} f * h_\lambda(x) = f(x).$$

Proof: Since $\int_{-\infty}^{\infty} h_\lambda(x) dx = 1$, we calculate

$$
\begin{aligned}
f * h_\lambda(x) - f(x) &= \int_{-\infty}^{\infty} f(y) h_\lambda(x - y) \, dy - \int_{-\infty}^{\infty} f(x) h_\lambda(y) \, dy \\
&= \int_{-\infty}^{\infty} (f(x - y) - f(x)) h_\lambda(y) dy \\
&= \int_{-\infty}^{\infty} (f(x - y) - f(x)) \frac{1}{\lambda} h_1(y/\lambda) dy \\
&= \int_{-\infty}^{\infty} (f(x - \lambda y) - f(x)) h_1(y) dy.
\end{aligned}
$$

If $|f(x)| \le C$ for all $x \in \mathbb{R}$, then the integrand is dominated by $2C h_1(y)$. As $\lambda \to 0$ we have that $f(x - \lambda y)$ tends to $f(x)$ locally uniformly in y. By the dominated convergence theorem we get the claim.   □

Theorem 3.4.4 *(Inversion formula) Let $f \in L^1_{bc}(\mathbb{R})$ and assume that $\hat{f}$ also lies in $L^1_{bc}(\mathbb{R})$, then for every $x \in \mathbb{R}$,*

$$\hat{\hat{f}}(x) = f(-x).$$

Another way to write this formula is

$$f(x) = \int_{-\infty}^{\infty} \hat{f}(y)e^{2\pi i x y}\, dy,$$

which means that f equals an integral over the plane waves $e^{2\pi i x y}$.

Proof: By Lemma 3.4.2 we have for $\lambda > 0$,

$$f * h_\lambda(x) = \int_{-\infty}^{\infty} e^{-\lambda|t|}\hat{f}(t)e^{2\pi i x t}\, dt.$$

The left-hand side tends to $f(x)$ as $\lambda \to 0$ by Lemma 3.4.3. The integrand on the right-hand side is dominated by $|\hat{f}(t)|$. The claim now follows by the dominated convergence theorem. $\qquad\square$

Corollary 3.4.5 *The Fourier transform restricted to S gives a bijection of the set S.*

Proof: Since S is mapped to itself, the corollary follows from the inversion theorem. $\qquad\square$

It will be useful later to have the following example at hand:

Proposition 3.4.6 *Let $f(x) = e^{-\pi x^2}$. Then $f \in S$ and*

$$\hat{f} = f.$$

Proof: According to Exercise 3.3 the function f is, up to scalar multiples, the unique solution of the differential equation

$$f'(x) = -2\pi x f(x).$$

By induction one deduces that for every natural number n there is a polynomial $p_n(x)$ such that $f^{(n)}(x) = p_n(x)e^{-\pi x^2}$. Since $e^{-\pi x^2}$ decreases faster than any power of x as $|x| \to \infty$, it follows that f lies in S. Then $\hat{f}$ also lies in S, and we compute

$$\begin{aligned}
(\hat{f})'(y) &= \int_{-\infty}^{\infty} (-2\pi i x)e^{-\pi x^2}e^{-2\pi i x y}\, dx \\
&= i\int_{-\infty}^{\infty} (e^{-\pi x^2})' e^{-2\pi i x y}\, dx \\
&= -2\pi y \hat{f}(y),
\end{aligned}$$

where we have used integration by parts. Thus we conclude that $\hat{f}(y) = ce^{-\pi y^2}$ for some constant c. Since $\hat{\hat{f}}(x) = f(-x)$, it follows that $c^2 = 1$, so $c = \pm 1$. Now $\hat{f}(0) = \int_{-\infty}^{\infty} e^{-\pi x^2}dx > 0$; therefore, $c = 1$. □

Corollary 3.4.7 *We have*

$$\int_{-\infty}^{\infty} e^{-x^2}dx = \sqrt{\pi}.$$

Proof: The proposition implies

$$\int_{-\infty}^{\infty} e^{-\pi x^2}dx = 1,$$

from which the corollary follows by a simple substitution. □

3.5   Plancherel's Theorem

Plancherel's theorem says that the Fourier transform preserves L^2-norms as follows. Let $L^2_{bc}(\mathbb{R})$ be the set of all continuous, bounded functions $f : \mathbb{R} \to \mathbb{C}$ with

$$\|f\|_2^2 \overset{\text{def}}{=} \int_{-\infty}^{\infty} |f(x)|^2 dx < \infty.$$

If f has this last property, we say that it is *square integrable*.

Lemma 3.5.1 *For any two functions* $f, g \in L^2_{bc}(\mathbb{R})$ *the integral*

$$\langle f, g \rangle = \int_{-\infty}^{\infty} f(x)\overline{g(x)}dx$$

converges and defines an inner product on the vector space $L^2_{bc}(\mathbb{R})$. *The space* $L^1_{bc}(\mathbb{R})$ *is a subspace of* $L^2_{bc}(\mathbb{R})$.

Proof: For $T > 0$ the space $C([-T, T])$ of continuous functions is a pre-Hilbert space with the inner product

$$\langle f, g \rangle_T = \int_{-T}^{T} f(x)\overline{g(x)}dx$$

(see Exercise 2.9). We write $\|.\|_{2,T}$ for the norm on this space. For $f, g \in L^2_{bc}(\mathbb{R})$ their restrictions to the interval $[-T, T]$ give elements of $C([-T, T])$, and the same holds for their absolute values $|f|$ and $|g|$.

Since the Cauchy inequality holds for elements of the vector space $C([-T, T])$, we may estimate

$$
\begin{aligned}
\int_{-T}^{T} |f(x)\overline{g(x)}| dx &= |\langle |f|, |g| \rangle_T| \leq \|f\|_{2,T} \|g\|_{2,T} \\
&= \sqrt{\int_{-T}^{T} |f(x)|^2 dx \int_{-T}^{T} |g(x)|^2 dx} \\
&\leq \sqrt{\int_{-\infty}^{\infty} |f(x)|^2 dx \int_{-\infty}^{\infty} |g(x)|^2 dx} \\
&= \|f\|_2 \|g\|_2 .
\end{aligned}
$$

Thus the integral is bounded by a constant not depending on T, which implies that the integral converges as T tends to infinity. The properties of an inner product are easily established.

For the last part let $f \in L^1_{bc}(\mathbb{R})$. Then f is bounded; say, $|f(x)| \leq C$ for every $x \in \mathbb{R}$. Then

$$|f(x)|^2 \leq C|f(x)|,$$

which implies that

$$\int_{-\infty}^{\infty} |f(x)|^2 dx \leq C \int_{-\infty}^{\infty} |f(x)| dx = C \|f\|_1 ,$$

and thus the former integral is finite, i.e., $f \in L^2_{bc}(\mathbb{R})$. □

By the lemma we see that $L^1_{bc}(\mathbb{R})$ is a pre-Hilbert space. We write $L^2(\mathbb{R})$ for its completion.

Theorem 3.5.2 *(Plancherel's theorem) For every $f \in L^1_{bc}(\mathbb{R})$ we have that $\hat{f} \in L^2_{bc}(\mathbb{R})$ and*

$$\|f\|_2 = \|\hat{f}\|_2.$$

In particular, the Fourier transform $f \mapsto \hat{f}$ extends to a unitary map $L^2(\mathbb{R}) \to L^2(\mathbb{R})$.

Proof: Let $\tilde{f}(x) = \overline{f(-x)}$ and let $g = \tilde{f} * f$. Then

$$g(x) = \int_{-\infty}^{\infty} \overline{f(y-x)} f(y) dy,$$

so that

$$g(0) = \|f\|_2^2.$$

Now $\hat{g}(t) = \hat{\tilde{f}}(t) \hat{f}(t) = \overline{\hat{f}(t)} \hat{f}(t) = |\hat{f}(t)|^2$. Therefore, we get

$$
\begin{aligned}
\|f\|_2^2 &= g(0) = \lim_{\lambda \to 0} g * h_\lambda(0) \\
&= \lim_{\lambda \to 0} \int_{-\infty}^{\infty} e^{-\lambda|t|} \hat{g}(t) dt \\
&= \lim_{\lambda \to 0} \int_{-\infty}^{\infty} e^{-\lambda|t|} |\hat{f}(t)|^2 dt = \|\hat{f}\|_2^2,
\end{aligned}
$$

by the monotone convergence theorem. $\square$

3.6   The Poisson Summation Formula

In this central section we bring together Fourier analysis on $\mathbb{R}$ and on $\mathbb{R}/\mathbb{Z}$ to derive the beautiful and expedient Poisson summation formula.

Let $f : \mathbb{R} \to \mathbb{C}$ be continuous and assume that for every $x \in \mathbb{R}$ the sum

$$g(x) = \sum_{l \in \mathbb{Z}} f(x + l)$$

converges absolutely. Then g defines a periodic function. Assume that its Fourier series converges pointwise to the function g; then

$$g(x) = \sum_{k \in \mathbb{Z}} c_k(g) e^{2\pi i k x},$$

so that for $x = 0$ we get

$$
\begin{aligned}
\sum_{l \in \mathbb{Z}} f(l) &= g(0) = \sum_{k \in \mathbb{Z}} c_k(g) \\
&= \sum_{k \in \mathbb{Z}} \int_0^1 g(y) e^{-2\pi i k y} dy \\
&= \sum_{k \in \mathbb{Z}} \int_0^1 \sum_{l \in \mathbb{Z}} f(y + l) e^{-2\pi i k y} dy.
\end{aligned}
$$

Assuming that we may interchange summation and integration, this equals

$$\sum_{k \in \mathbb{Z}} \sum_{l \in \mathbb{Z}} \int_l^{l+1} f(y) e^{-2\pi i k y} dy = \sum_{k \in \mathbb{Z}} \int_{-\infty}^{\infty} f(y) e^{-2\pi i k y} dy = \sum_{k \in \mathbb{Z}} \hat{f}(k).$$

This is a formal computation, valid only under certain assumptions. We will now turn it into a theorem by giving a set of conditions that ensures the validity of those assumptions.

Theorem 3.6.1 Let $f \in L^1_{\text{bc}}(\mathbb{R})$ and assume that f is piecewise continuously differentiable with the possible exception of finitely many points. Let

$$\varphi(x) = \begin{cases} f'(x) & \text{if it exists,} \\ 0 & \text{otherwise,} \end{cases}$$

and assume that $x^2 f(x)$ and $x^2 \varphi(x)$ are bounded. Then the Poisson summation formula

$$\sum_{k \in \mathbb{Z}} f(k) = \sum_{k \in \mathbb{Z}} \hat{f}(k)$$

holds.

Proof: Since $x^2 f(x)$ is bounded, the series $g(x) = \sum_{k \in \mathbb{Z}} f(x + k)$ converges uniformly to give a continuous function g. Likewise, the sum $\sum_{k \in \mathbb{Z}} \varphi(x + k)$ converges to a piecewise continuous function $\tilde{g}$. Since $f(x) = c + \int_0^x \varphi(t) dt$, it follows that

$$\int_0^x \tilde{g}(t) dt = \int_0^x \sum_{k \in \mathbb{Z}} \varphi(t + k) dt = \sum_{k \in \mathbb{Z}} \int_0^x \varphi(t + k) dt$$

$$= \sum_{k \in \mathbb{Z}} \int_k^{k+x} \varphi(t) dt = \sum_{k \in \mathbb{Z}} f(k + x) - f(k)$$

$$= g(x) - g(0),$$

where we were allowed to interchange integration and summation because the sum converges uniformly. It follows that g is piecewise continuously differentiable, and so the Fourier series converges pointwise.

The integration and summation may be interchanged, since the sum converges uniformly on the interval $[0, 1]$. This finishes the proof of the Poisson summation formula. $\qquad \square$

3.7   Theta Series

As an application of the Poisson summation formula we give a proof
of the functional equation of the classical theta series. In Appendix
A this is employed to derive the analytic continuation and the functional equation of the Riemann zeta function. Since this is a result
of utmost importance to many areas of mathematics, it is included
in this book. It requires, however, knowledge of complex analysis,
which is why it appears only in the appendix.

Theorem 3.7.1 *For $t > 0$ let*

$$\Theta(t) = \sum_{k \in \mathbb{Z}} e^{-t\pi k^2}.$$

Then for every $t > 0$ we have

$$\Theta(t) = \frac{1}{\sqrt{t}} \Theta\left(\frac{1}{t}\right).$$

Proof: Let $f_t(x) = e^{-t\pi x^2}$. Then by Proposition 3.4.6 we have
$\hat{f_1} = f_1$, and therefore by Theorem 3.3.1 d),

$$\hat{f_t}(x) = \frac{1}{\sqrt{t}} f_{1/t}(x).$$

Since f_t is in $\mathcal{S}$, Theorem 3.6.1 applies to give

$$\sum_{k \in \mathbb{Z}} f_t(k) = \sum_{k \in \mathbb{Z}} \hat{f_t}(k).$$

$\square$

3.8   Exercises

Exercise 3.1 Show that

$$\int_{-\infty}^{\infty} \frac{1}{1+x^2} dx = \pi.$$

Exercise 3.2 Let $a < b$ be real numbers and let

$$g(x) = \begin{cases} 1 & \text{if } a \leq x \leq b, \\ 0 & \text{otherwise.} \end{cases}$$

Compute the Fourier transform $\hat{g}(x)$.

Exercise 3.3 Let $g : \mathbb{R} \to \mathbb{C}$ be continuously differentiable and satisfy the differential equation

$$g'(x) = -2\pi x g(x).$$

Show that there is a constant c such that $g(x) = c e^{-\pi x^2}$.

(Hint: Set $u(x) = g(x) e^{\pi x^2}$ and deduce that $u'(x) = 0$.)

Exercise 3.4 Let $D = \frac{d}{dx}$ be the ordinary differential operator on $\mathbb{R}$. Let f and g be n-times continuously differentiable on $\mathbb{R}$. Show that

$$D^n(fg) = \sum_{k=0}^{n} \binom{n}{k} D^k g D^{n-k} f.$$

Exercise 3.5 Let $f(x) = e^{-x^2}$. Prove that for every $n \geq 0$ there is a polynomial $p_n(x)$ such that $D^n f(x) = p_n(x) f(x)$, and conclude from this that $f(x)$ lies in $\mathcal{S}$.

Exercise 3.6 Let $f(x) = e^{-x^2}$. Compute $f * f$.

Exercise 3.7 Let $f \in L^1_{bc}(\mathbb{R})$, $f > 0$. Prove that $\left| \hat{f}(y) \right| < \hat{f}(0)$ for every $y \neq 0$.

Exercise 3.8 A function f on $\mathbb{R}$ is called locally integrable if f is integrable on every bounded interval $[a, b]$ for $a < b$ in $\mathbb{R}$. Show that if $g \in C_c^\infty(\mathbb{R})$ and f is locally integrable, then $f * g$ exists and is infinitely differentiable on $\mathbb{R}$.

Exercise 3.9 Show that for every $T > 0$ there is a smooth function with compact support $\chi : \mathbb{R} \to [0, 1]$ such that $\chi \equiv 1$ on $[-T, T]$.

(Hint: Choose a suitable $g \in C_c^\infty(\mathbb{R})$ and consider $f * g$, where f is the characteristic function of some interval.)

Exercise 3.10 Show that a sequence (f_n) of functions on $\mathbb{R}$ converges locally uniformly to the function f if and only if it converges uniformly on every bounded interval $[a, b]$, $a, b \in \mathbb{R}$, $a < b$.

Exercise 3.11 Prove that for $a > 0$ the following holds:

$$\frac{1 + e^{-2\pi a}}{1 - e^{-2\pi a}} = \frac{1}{\pi} \sum_{n=-\infty}^{\infty} \frac{a}{a^2 + n^2}.$$

(Hint: Apply the Poisson summation formula to $f(x) = e^{-2\pi a |x|}$.)

Chapter 4

Distributions

We have seen that the Fourier transform extends to a unitary map from $L^2(\mathbb{R})$ to itself. However, it is also defined for functions that are unbounded and integrable but not in $L^2(\mathbb{R})$. So the question arises as to the ultimate domain of definition for the Fourier transform that contains all these spaces. A possible answer will be given in this chapter, since we will see that the Fourier transform extends nicely to the space of tempered distributions.

4.1 Definition

Let $C^\infty(\mathbb{R})$ be the vector space of all infinitely differentiable functions on $\mathbb{R}$. We say that a function $f\colon \mathbb{R} \to \mathbb{C}$ has *compact support* if $f \equiv 0$ outside a bounded interval. Let $C_c^\infty(\mathbb{R})$ be the complex vector space of all infinitely differentiable functions on $\mathbb{R}$ with values in $\mathbb{C}$ that have compact support. It is not a priori clear that this space it not the zero space. So we give a construction of a nonzero element. Let

$$f(x) = \begin{cases} 0, & x \leq 0, \\ e^{-\frac{1}{x}} e^{-\frac{1}{1-x}}, & 0 < x < 1, \\ 0, & x \geq 1. \end{cases}$$

Then f is a function that is zero outside the interval $[0,1]$. To see that it actually is smooth, it suffices to show that the function

$$h(x) = \begin{cases} 0, & x \leq 0, \\ e^{-\frac{1}{x}}, & x > 0, \end{cases}$$

is infinitely differentiable. For this, one has only to consider the point $x_0 = 0$, since h is clearly smooth everywhere else. One finds that any derivative $h^{(n)}(x)$, $x \neq 0$, tends to zero as $x \to 0$. This implies the smoothness at $x_0 = 0$.

Having one nonzero function f in $C_c^\infty(\mathbb{R})$, we now can take linear combinations and products of functions of the form $f(ax + b)$, $a \neq 0$, to see that $C_c^\infty(\mathbb{R})$ actually is a rather large space.

We say that a sequence $(g_n)_{n\in\mathbb{N}}$ in $C_c^\infty(\mathbb{R})$ converges[1] to $g \in C_c^\infty(\mathbb{R})$ if there is a bounded interval I such that $g_n \equiv 0$ outside I for every n, and every derivative $(g_n^{(k)})_{n\in\mathbb{N}}$ converges uniformly to $g^{(k)}$.

A *distribution* on $\mathbb{R}$ is a linear map

$$T \colon C_c^\infty(\mathbb{R}) \to \mathbb{C}$$

such that

$$\lim_{n\to\infty} T(g_n) \;=\; T(g)$$

whenever g_n converges to g in $C_c^\infty(\mathbb{R})$.

Examples.

- The *delta distribution*

$$\delta(f) \;\overset{\text{def}}{=}\; f(0),$$

  also called *Dirac distribution* or *Dirac delta*.

- The integral

$$I(f) \;\overset{\text{def}}{=}\; \int_{-\infty}^{\infty} f(x)\,dx.$$

- A function ϕ on $\mathbb{R}$ is called *locally Riemann integrable* if it is Riemann integrable on every bounded interval. A given locally Riemann integrable function ϕ defines a distribution I_ϕ by

$$I_\phi(f) \;\overset{\text{def}}{=}\; \int_{-\infty}^{\infty} f(x)\,\phi(x)\,dx.$$

---

[1] The reader should be aware that this notion does not fully describe the topology on $C_c^\infty(\mathbb{R})$, which is an inductive limit topology. As long as one considers only linear maps to $\mathbb{C}$, however, it suffices to consider sequences as given here.

We denote the complex vector space of all distributions on $\mathbb{R}$ by $C_c^\infty(\mathbb{R})'$. Motivated by the mapping $\phi \mapsto I_\phi$, we sometimes call distributions *generalized functions*. A distribution T in general is not a function, so it does not make sense to write $T(x)$, but nevertheless, it is convenient to write

$$T(f) = \int_{-\infty}^{\infty} T(x)f(x)\,dx.$$

For instance, the distribution $T(x-a)$, $a \in \mathbb{R}$, is defined by

$$\int_{-\infty}^{\infty} T(x-a)f(x)\,dx \overset{\text{def}}{=} \int_{-\infty}^{\infty} T(x)f(x+a)\,dx.$$

So $T(x-a)$ applied to f equals T applied to $x \mapsto f(x+a)$. For example,

$$\int_{-\infty}^{\infty} \delta(x-a)f(x)\,dx = \int_{-\infty}^{\infty} \delta(x)f(x+a)\,dx = f(a).$$

Via the map $\phi \mapsto I_\phi$ we can also identify the space $L_{bc}^2(\mathbb{R})$ with a subspace of $C_c^\infty(\mathbb{R})'$.

Many authors use the symbol $\mathcal{D}$ for $C_c^\infty(\mathbb{R})$ and $\mathcal{D}'$ for $C_c^\infty(\mathbb{R})'$.

4.2 The Derivative of a Distribution

Distributions have many nice analytic properties. We will illustrate this by the fact that unlike functions, distributions are always differentiable. This is motivated as follows.

Let ϕ be a continuously differentiable function on $\mathbb{R}$. Integration by parts implies that for $g \in C_c^\infty(\mathbb{R})$,

$$I_{\phi'}(g) = \int_{\mathbb{R}} \phi'(x)g(x)\,dx = -\int_{\mathbb{R}} \phi(x)g'(x)\,dx = -I_\phi(g').$$

This opens the way to generalize differentiation to the space of distributions as follows. Let T be a distribution. We define its *derivative* $T' \in C_c^\infty(\mathbb{R})'$ by

$$T'(g) \overset{\text{def}}{=} -T(g').$$

As an example, consider a function ϕ that is continuously differentiable; then $I'_\phi = I_{\phi'}$. As another example, let ϕ be the characteristic function of the unit interval $[0,1]$. Then for $g \in C_c^\infty(\mathbb{R})$,

$$I'_\phi(g) = -T_\phi(g') = -\int_0^1 g'(x)\,dx = g(0) - g(1),$$

so that we can write $I'_\phi(x) = \delta(x) - \delta(x-1)$.

4.3   Tempered Distributions

Recall the space of Schwartz functions $\mathcal{S} = \mathcal{S}(\mathbb{R})$ consisting of all C^∞ functions f on $\mathbb{R}$ such that for all $m, n \geq 0$ we have

$$\sigma_{m,n}(f) = \sup_{x \in \mathbb{R}} |x^m f^{(n)}(x)| < \infty.$$

By Proposition 3.3.3 we know that $\mathcal{S}$ is stable under the Fourier transform. We say that a sequence $(f_j)_{j \in \mathbb{N}}$ in $\mathcal{S}$ converges to $f \in \mathcal{S}$ if for every pair of integers $m, n \geq 0$ the numbers $\sigma_{m,n}(f_j - f)$ tend to zero.

Lemma 4.3.1 *The space $\mathcal{S}$ is a subspace of $L_{bc}^2(\mathbb{R})$, and for every sequence (f_j) in $\mathcal{S}$ that converges to $f \in \mathcal{S}$ we have $\lim_j \|f_j - f\|_2 = 0$.*

Proof: Suppose $f_j \to f$ in $\mathcal{S}$. Then in particular the sequences $\sigma_{0,0}(f_j - f)$ and $\sigma_{1,0}(f_j - f)$ tend to zero. Set $C = \int_{-\infty}^\infty \frac{1}{(1+|x|)^2}\,dx$ and let $\varepsilon > 0$. Then there is $j_0 \in \mathbb{N}$ such that for all $j \geq j_0$,

$$\sup_{x \in \mathbb{R}} |f_j(x) - f(x)|(1+|x|) = \sigma_{0,0}(f_j - f) + \sigma_{1,0}(f_j - f) < \sqrt{\varepsilon/C}.$$

Let $j \geq j_0$. Then $|f_j(x) - f(x)| < \frac{\sqrt{\varepsilon/C}}{1+|x|}$ for every $x \in \mathbb{R}$. This implies

$$\|f_j - f\|_2^2 = \int_{-\infty}^\infty |f_j(x) - f(x)|^2\,dx < \int_{-\infty}^\infty \frac{\varepsilon/C}{(1+|x|)^2}\,dx = \varepsilon.$$

$\square$

A *tempered distribution* is a linear map

$$T\colon \mathcal{S} \to \mathbb{C}$$

such that

$$\lim_{k \to \infty} T(f_k) = T(f)$$

for every convergent sequence (f_k) in $\mathcal{S}$ with limit f. It is easy to see that if $g_j \in C_c^\infty(\mathbb{R})$ converges to g in $C_c^\infty(\mathbb{R})$, then $T(g_j) \to T(g)$ for every tempered distribution T, and so for every tempered distribution T the restriction $T|_{C_c^\infty(\mathbb{R})}$ is a distribution. The space of all tempered distributions is denoted by $\mathcal{S}'$.

Proposition 4.3.2 *The restriction*

$$\mathcal{S}' \to C_c^\infty(\mathbb{R})',$$
$$T \mapsto T|_{C_c^\infty(\mathbb{R})},$$

is injective. So we can consider the space of tempered distributions as a subspace of the space of all distributions.

Proof: Let T be a tempered distribution with $T|_{C_c^\infty(\mathbb{R})} = 0$. Let $f \in \mathcal{S}$. We have to show that $T(f) = 0$. For this let η be a smooth function from $\mathbb{R}$ to the unit interval $[0, 1]$ with $\eta(x) = 0$ for $x \leq 0$ and $\eta(x) = 1$ for $x \geq 1$. For $n \in \mathbb{N}$ set

$$\chi_n(x) \stackrel{\text{def}}{=} \eta(n + x)\eta(n - x).$$

Then χ_n has compact support and $\chi_n(x) = 1$ for $|x| \leq n - 1$. Set $f_n(x) = \chi_n(x)f(x)$. Then each f_n lies in $C_c^\infty(\mathbb{R})$, and the sequence f_n converges to f in $\mathcal{S}$. (See Exercise 6.6.) Hence $T(f) = \lim_n T(f_n) = 0$. $\qquad\qquad\square$

Let ϕ be a locally integrable function on $\mathbb{R}$. Then under mild restrictions the distribution I_ϕ extends to a tempered distribution, as the following lemma shows.

Lemma 4.3.3 *Let ϕ be a locally integrable function on $\mathbb{R}$ such that there is a natural number n with*

$$\int_{-\infty}^{\infty} |\phi(x)| \frac{1}{1 + x^{2n}} \, dx < \infty.$$

Then the integral $I_\phi(f) = \int_{-\infty}^{\infty} \phi(x)f(x) \, dx$ converges for every $f \in \mathcal{S}$ and defines a tempered distribution $f \mapsto I_\phi(f)$.

Proof: The convergence of the integral is clear, so that it remains to show that it indeed defines a tempered distribution. Let ϕ and n be as in the lemma. Let $C = \int_{-\infty}^{\infty} |\phi(x)| \frac{1}{1+x^{2n}}\, dx$ and assume without loss of generality that $C > 0$. Suppose the sequence (f_k) converges to f in $\mathcal{S}$. Let $\varepsilon > 0$. Then there is k_0 such that for $k \geq k_0$ we have $\sup_{x \in \mathbb{R}} |f_k(x) - f(x)| < \frac{\varepsilon}{C(1+x^{2n})}$. Then, for $k \geq k_0$,

$$\begin{aligned}
|I_\phi(f_k) - I_\phi(f)| &\leq \int_{-\infty}^{\infty} |\phi(x)|\,|f_k(x) - f(x)|\, dx \\
&< \frac{\varepsilon}{C} \int_{-\infty}^{\infty} \frac{|\phi(x)|}{1 + x^{2n}} dx = \varepsilon.
\end{aligned}$$

$\square$

Proposition 4.3.4 *The map $\phi \mapsto I_\phi$ from $L^2_{bc}(\mathbb{R})$ to $\mathcal{S}'$ is injective and extends to a natural embedding $L^2(\mathbb{R}) \hookrightarrow \mathcal{S}'$.*

Proof: Let $\phi \in L^2_{bc}(\mathbb{R})$ with $I_\phi = 0$. For $f \in \mathcal{S}$ we then have $0 = I_\phi(\bar{f}) = \langle \phi, f \rangle$, where $\bar{f}(x) = \overline{f(x)}$ is the complex conjugate. Since $\mathcal{S}$ is dense in $L^2(\mathbb{R})$, this implies that $\phi = 0$, and thus we have established the injectivity of the map $\phi \mapsto I_\phi$.

To see that it extends, let $(\phi_n)_{n \in \mathbb{N}}$ be a Cauchy sequence in $L^2_{bc}(\mathbb{R})$ with limit $\phi \in L^2(\mathbb{R})$. Let $f \in \mathcal{S}$. Since $\mathcal{S}$ is a subset of $L^2_{bc}(\mathbb{R})$, we can write

$$I_{\phi_n}(f) = \int_{\mathbb{R}} \phi_n(x) f(x)\, dx = \langle \phi_n, \bar{f} \rangle.$$

For $m, n \in \mathbb{N}$ we have

$$|I_{\phi_n}(f) - I_{\phi_m}(f)| = |\langle \phi_n - \phi_m, \bar{f} \rangle| \leq \|\phi_n - \phi_m\|_2 \, \|f\|_2 .$$

This implies that $I_{\phi_n}(f)$ is a Cauchy sequence. Define

$$I_\phi(f) \stackrel{\text{def}}{=} \lim_{n \to \infty} I_{\phi_n}(f).$$

Since this works for any sequence with limit ϕ, it follows that this definition does not depend on the choice of the sequence ϕ_n. It finally remains to show that the limit obtained indeed is a distribution, i.e., satisfies the requirement of continuity. So suppose (f_j) converges to

f in $\mathcal{S}$. We compute

$$
\begin{aligned}
|I_\phi(f_j) - I_\phi(f)| &= |I_\phi(f_j - f)| \\
&= \lim_n |I_{\phi_n}(f_j - f)| \\
&= \lim_n |\langle \phi_n, \overline{f_j - f} \rangle| \\
&\leq \left(\lim_n \|\phi_n\|_2 \right) \|f_j - f\|_2 \\
&= \|\phi\|_2 \, \|f_j - f\|_2 .
\end{aligned}
$$

The latter tends to zero by Lemma 4.3.1. This implies that $I_\phi(f_j)$ converges to $I_\phi(f)$ as $j \to \infty$, so I_ϕ is indeed a tempered distribution. $\qquad\square$

4.4   Fourier Transform

For $f \in \mathcal{S}$ we write $f^\vee(x) = f(-x)$. Then the Fourier inversion theorem tells us that $\hat{\hat{f}} = f^\vee$. Now for $\phi, f \in \mathcal{S}$ we have

$$
I_{\hat{\phi}}(f) = \left\langle \hat{\phi}, \overline{f} \right\rangle = \left\langle \hat{\hat{\phi}}, \hat{\overline{f}} \right\rangle = \left\langle \phi^\vee, \hat{\overline{f}} \right\rangle = \left\langle \phi, \hat{\overline{f}}^\vee \right\rangle .
$$

Further,

$$
\hat{\overline{f}}(x) = \int_{-\infty}^\infty \overline{f(y)} e^{-2\pi i x y} \, dy = \overline{\int_\infty^\infty f(y) e^{2\pi i x y} dy} = \overline{\hat{f}(-x)}.
$$

We conclude that $\hat{\overline{f}}^\vee = \overline{\hat{f}}$ and so

$$
I_{\hat{\phi}}(f) = I_\phi(\hat{f}).
$$

Again we turn this into a definition for all tempered distributions. We define the *Fourier transform* of a tempered distribution T by

$$
\hat{T}(f) \overset{\mathrm{def}}{=} T(\hat{f}).
$$

Examples.

- Let $I(f) = \int_\infty^\infty f(x) \, dx$. Then

$$
\hat{I}(f) = \int_{-\infty}^\infty \hat{f}(x) \, dx = \int_{-\infty}^\infty \hat{f}(x) \, dx = f(0).
$$

  So we get $\hat{I} = \delta$.

- Likewise, we compute the Fourier transform of the delta distribution to be

$$\hat{\delta}(f) = \delta(\hat{f}) = \hat{f}(0) = \int_{-\infty}^{\infty} f(x)\,dx = I(f).$$

The following lemma shows that there is a systematic reason for this.

Lemma 4.4.1 *For every tempered distribution T and every $f \in S$ we have*

$$\hat{T}(f) = T(f^{\vee}).$$

Proof: We have

$$\hat{T}(f) = \hat{T}(\hat{f}^{\vee}) = T((\widehat{\hat{f}^{\vee}})^{\vee}).$$

Now

$$\begin{aligned}
(\widehat{\hat{f}^{\vee}})^{\vee}(x) &= (\widehat{\hat{f}^{\vee}})(x) \\
&= \int_{-\infty}^{\infty} \hat{f}^{\vee}(y) e^{2\pi i x y}\,dy \\
&= \int_{-\infty}^{\infty} \hat{f}(y) e^{-2\pi i x y}\,dy \\
&= \hat{\hat{f}}(x) = f(-x) = f^{\vee}(x).
\end{aligned}$$

We say that a smooth function f has *moderate growth* if for every $k \geq 0$ there is $l \in \mathbb{N}$ such that

$$\sup_{x \in \mathbb{R}} \frac{|f^{(k)}(x)|}{1 + x^{2l}} < \infty.$$

In other words, this means that every derivative of f grows at most like a polynomial.

Lemma 4.4.2 *Let f be of moderate growth and $g \in S$. Then the pointwise product fg lies in S again. If the sequence (g_j) converges to g in S, then fg_j converges to fg.*

Proof: Let $m, n \geq 0$. Then

$$
\begin{aligned}
\left| x^m (fg)^{(n)}(x) \right| &= \left| x^m \sum_{k=0}^{n} \binom{n}{k} f^{(k)}(x) g^{(n-k)}(x) \right| \\
&\leq |x|^m \sum_{k=0}^{n} \binom{n}{k} |f^{(k)}(x)| \, |g^{(n-k)}(x)| \\
&\leq C |x|^m (1 + x^{2M}) \sum_{k=0}^{n} \binom{n}{k} |g^{(n-k)}(x)|
\end{aligned}
$$

for some $C > 0$, $M \in \mathbb{N}$ depending on f and n. Therefore, we get

$$
\sigma_{m,n}(fg) \leq C \sum_{k=0}^{n} \binom{n}{k} \left(\sigma_{m,n-k}(g) - \sigma_{m+2M,n-k}(g) \right).
$$

It follows that $fg \in \mathcal{S}$. Next suppose that the sequence (g_j) converges to g in $\mathcal{S}$. Let $m, n \geq 0$ and let $\varepsilon > 0$. Then there is $j_0 \in \mathbb{N}$ such that for $j \geq j_0$,

$$
C \sum_{k=0}^{n} \binom{n}{k} \left(\sigma_{m,n-k}(g_j - g) - \sigma_{m+2M,n-k}(g_j - g) \right) < \varepsilon.
$$

This implies that

$$
\sigma_{m,n}(fg_j - fg) < \varepsilon
$$

for $j \geq j_0$. $\qquad\square$

For a function f of moderate growth and a tempered distribution T we define their product fT by

$$
fT(g) \overset{\text{def}}{=} T(fg)
$$

for $g \in \mathcal{S}$.

Theorem 4.4.3 *Let T be a tempered distribution.*

(a)   If $S(x) = -2\pi i x T(x)$, then $(\hat{T})' = \hat{S}$.

(b) $\widehat{(T')}(y) = 2\pi i y \hat{T}(y)$.

Proof: Let $S(x) = -2\pi i x T(x)$. Then for $f \in \mathcal{S}$,

$$
\begin{aligned}
(\hat{T})'(f) &= -\hat{T}(f') = -T(\hat{f}') \\
&= -T(2\pi i y \hat{f}) \\
&= S(\hat{f}) = \hat{S}(f).
\end{aligned}
$$

For the second part compute

$$
\begin{aligned}
\widehat{T'}(f) &= T'(\hat{f}) = -T(\hat{f}') \\
&= T(\widehat{2\pi i x f(x)}) = \hat{T}(2\pi i x f(x)) \\
&= 2\pi i y \hat{T}(f).
\end{aligned}
$$

$\square$

4.5 Exercises

Exercise 4.1 Show that the Hilbert space completion $L^2(\mathbb{R})$ of $L^2_{bc}(\mathbb{R})$ is also the Hilbert space completion of the space $C_c^\infty(\mathbb{R})$ of all infinitely differentiable functions with compact support.

Exercise 4.2 Let T be a tempered distribution and let $S(x) = e^{2\pi i a x} T(x)$ for some $a \in \mathbb{R}$. Show that $\hat{S}(y) = T(y - a)$.

Exercise 4.3 Let T be a tempered distribution and let $S(x) - T(x - a)$. Show that $\hat{S}(y) = e^{-2\pi i a y} T(y)$.

Exercise 4.4 Show that $C_c^\infty(\mathbb{R})$ is dense in $\mathcal{S}$. More precisely, show that for every $f \in \mathcal{S}$ the sequence $f_n = \chi_n f$ as in the proof of Proposition 4.3.2 converges to f in $\mathcal{S}$.

Exercise 4.5 For $f \in \mathcal{S}$ let

$$
T(f) \stackrel{\text{def}}{=} \sum_{k \in \mathbb{Z}} f(k).
$$

Show that T is a tempered distribution and that $\hat{T} = T$.

Exercise 4.6 A distribution T is said to have *compact support* if there is a bounded interval I such that $T(g) = 0$ for every g that vanishes on I. Choose $\psi \in C_c^\infty(\mathbb{R})$ with $\psi \equiv 1$ on I.

(a)   Show that the linear map

$$C^\infty(\mathbb{R}) \;\to\; \mathbb{C},$$
$$f \;\mapsto\; T(\psi f),$$

does not depend on the choice of I or ψ. By abuse of notation this linear map is also called T.

(b)   Let T be a distribution with compact support. Show that the Fourier transform of T is a function. More precisely,

$$\hat{T}(x) \;=\; T_y(e^{-2\pi i x y}),$$

where the notation T_y indicates that T is applied to the function of the variable y.

Part II

LCA Groups

Chapter 5

Finite Abelian Groups

In this chapter we present the complete theory developed in this book for the simplest case to which it can be applied, that of a finite abelian group. In this case no analytic tools are required, and only a small amount of group theory is needed in order to understand the concept of the duality and the Plancherel theorem.

5.1   The Dual Group

Let A be a finite abelian group. The group A is called *cyclic* if it is generated by a single element, i.e., if there is $\tau \in A$, called a *generator* of A, such that $A = \{1, \tau, \tau^2, \ldots, \tau^{N-1}\}$, where $N = |A|$ is the cardinality of the set A. We will make use of the following result.

Theorem 5.1.1 *(Main theorem on finite abelian groups) Any finite abelian group A is isomorphic to a product $A_1 \times A_2 \times \cdots \times A_k$ of cyclic groups.*

Proof: [17], Theorem 10.1.   □

Let A be a finite abelian group. A *character* χ of A is a group homomorphism $\chi : A \to \mathbb{T}$ to the unit torus, so χ is a map satisfying

$$\chi(ab) \;=\; \chi(a)\chi(b)$$

for every $a, b \in A$. Let $\hat{A}$ be the set of all characters of A.

Lemma 5.1.2 *The pointwise product* $(\chi, \eta) \mapsto \chi\eta$ *with*

$$\chi\eta(a) \;=\; \chi(a)\eta(a)$$

makes $\hat{A}$ *an abelian group. We call* $\hat{A}$ *the dual group, or Pontryagin dual, of* A.

Proof: We have to show that $\chi\eta$ is a character when χ and η are. For this we let $a, b \in A$ and compute

$$\begin{aligned}
\chi\eta(ab) \;=\;& \chi(ab)\eta(ab) \;=\; \chi(a)\chi(b)\eta(a)\eta(b) \\
=\;& \chi(a)\eta(a)\chi(b)\eta(b) \;=\; \chi\eta(a)\,\chi\eta(b).
\end{aligned}$$

In the same way we establish that χ^{-1} is a character when χ is one, where $\chi^{-1}(a) = \chi(a)^{-1}$. This shows that $\hat{A}$ is a subgroup of the group of all maps from A to $\mathbb{T}$. $\qquad\square$

Lemma 5.1.3 *Let* A *be cyclic of order* N. *Fix a generator* τ *of* A, *i.e.,* $A = \left\{ 1, \tau, \tau^2, \ldots, \tau^{N-1} \right\}$ *and* $\tau^N = 1$. *The characters of the group* A *are given by*

$$\eta_l(\tau^k) \;=\; e^{2\pi i k l/N}, \quad k \in \mathbb{Z},$$

for $l = 0, 1, \ldots, N-1$. *The group* $\hat{A}$ *is again cyclic of order* N.

Proof: Let η be a character of A. Then $\eta(\tau)$ is an element t of $\mathbb{T}$ that satisfies $t^N = \eta(\tau^N) = 1$. Therefore, there is a unique $l \in \{0, 1, \ldots, N-1\}$ such that

$$\eta(\tau) \;=\; e^{2\pi i l/N}.$$

For every $k \in \mathbb{Z}$ we get

$$\eta(\tau^k) \;=\; \eta(\tau)^k \;=\; e^{2\pi i k l/N}.$$

This shows that every character is of the form η_l for some $l \in \{0, 1, \ldots, N-1\}$. It is easy to see that $\eta_l \neq \eta_{l'}$ if $l \neq l'$, so the lemma follows. $\qquad\square$

We conclude that for every finite cyclic group A its dual $\hat{A}$ is also a cyclic group of the same order. This then implies that those two groups must be isomorphic. In general, there will be several isomorphisms and no naturally preselected, or canonical, one. One gets a canonical isomorphism when one goes one step further; i.e., one considers the bidual, which is the dual of the dual.

Theorem 5.1.4 *Let A be a finite abelian group. There is a canonical isomorphism to the bidual $A \to \hat{\hat{A}}$ given by $a \mapsto \delta_a$, where δ_a is the point evaluation at a, i.e.,*

$$\delta_a \; : \; \hat{A} \; \to \; \mathbb{T},$$
$$\chi \; \mapsto \; \chi(a).$$

Proof: The map $a \mapsto \delta_a$ is a homomorphism, since

$$\delta_{ab}(\chi) \; = \; \chi(ab) \; = \; \chi(a)\chi(b) \; = \; \delta_a(\chi)\delta_b(\chi).$$

Moreover, the following lemma shows that this map is injective.

Lemma 5.1.5 *Let A be a finite abelian group and let $a \in A$. Suppose that $\chi(a) = 1$ for every $\chi \in \hat{A}$. Then $a = 1$.*

Proof: Lemma 5.1.3 shows that the claim holds for cyclic groups. In the light of the main theorem on finite abelian groups it remains to show that if the claim holds for the groups A and B, then it also holds for $A \times B$. For this let $(a_0, b_0) \in A \times B$ with $\eta(a_0, b_0) = 1$ for all $\eta \in \widehat{A \times B}$. For every $\chi \in \hat{A}$ the map $\chi(a, b) = \chi(a)$ is a character of $A \times B$, and therefore $\chi(a_0) = 1$, which implies $a_0 = 1$ and similarly $b_0 = 1$. The lemma is proven.

The lemma implies that the map $a \mapsto \delta_a$ is injective from A to $\hat{\hat{A}}$. Since the cardinality $|A|$ is the same as the cardinality $|\hat{A}|$ of the dual (Exercise 5.2), and by iteration the same as the cardinality of the bidual, the theorem follows. $\qquad\square$

5.2 The Fourier Transform

Let A be a finite abelian group. The Hilbert space $\ell^2(A)$ coincides with the space $\mathbb{C}^A$ of all maps from A to $\mathbb{C}$. In particular, the characters $\chi : A \to \mathbb{T} \subset \mathbb{C}$ are elements of $\ell^2(A)$.

Lemma 5.2.1 *Let χ, η be characters of A; then*

$$\langle \chi, \eta \rangle \; = \; \begin{cases} |A| & \text{if } \chi = \eta, \\ 0 & \text{otherwise.} \end{cases}$$

Proof: First consider the case where $\chi = \eta$; then

$$\langle \chi, \eta \rangle = \sum_{a \in A} \chi(a)\overline{\eta(a)} = \sum_{a \in A} |\chi(a)|^2 = \sum_{a \in A} 1 = |A|.$$

Next assume $\chi \neq \eta$; then the character $\alpha = \chi\eta^{-1}$ is different from 1 and

$$\langle \chi, \eta \rangle = \sum_{a \in A} \chi(a)\eta(a)^{-1} = \sum_{a \in A} \alpha(a).$$

Let $b \in A$ with $\alpha(b) \neq 1$. Then

$$\langle \chi, \eta \rangle \alpha(b) = \sum_{a \in A} \alpha(a)\alpha(b) = \sum_{a \in A} \alpha(ab).$$

Replacing the sum index a by ab^{-1}, which also runs over the entire group, we see that this yields

$$\sum_{a \in A} \alpha(ab) = \sum_{a \in A} \alpha(a) = \langle \chi, \eta \rangle.$$

Thus $(\alpha(b) - 1)\langle \chi, \eta \rangle = 0$, which implies $\langle \chi, \eta \rangle = 0$. $\square$

For $f \in \ell^2(A)$ we define its *Fourier transform* $\hat{f} : \hat{A} \to \mathbb{C}$ by

$$\hat{f}(\chi) = \frac{1}{\sqrt{|A|}} \langle f, \chi \rangle = \frac{1}{\sqrt{|A|}} \sum_{a \in A} f(a)\overline{\chi(a)}.$$

The presence of the normalizing factor $1/\sqrt{|A|}$ needs explanation, in particular since no such factor shows up in the Fourier transform on $\mathbb{R}$. This factor is needed here because of the normalization of the inner products on $\ell^2(A)$ and $\ell^2(\hat{A})$. In the case of the Fourier transform on $\mathbb{R}$ no such factor appeared, since by writing the characters of $\mathbb{R}$ as $e^{2\pi i x y}$ instead of e^{ixy} we already implicitly included a normalizing factor of 2π in the inner product on the dual. Without this normalization the Fourier transform over $\mathbb{R}$ would have had a normalizing factor of $1/\sqrt{2\pi}$.

Theorem 5.2.2 *The map $f \mapsto \hat{f}$ is an isomorphism of the Hilbert spaces $\ell^2(A) \to \ell^2(\hat{A})$. This can also be applied to the group $\hat{A}$, and the composition of the two Fourier transforms gives a map $f \mapsto \hat{\hat{f}}$. For the latter map we have*

$$\hat{\hat{f}}(\delta_a) = f(a^{-1}).$$

Proof: Let $f, g \in \ell^2(A)$. We have to show that $\left\langle \hat{f}, \hat{g} \right\rangle = \langle f, g \rangle$, where the inner products are taken on $\hat{A}$ and A respectively. For this we compute

$$
\begin{aligned}
\left\langle \hat{f}, \hat{g} \right\rangle &= \sum_{\chi \in \hat{A}} \hat{f}(\chi)\overline{\hat{g}(\chi)} \\
&= \frac{1}{|A|} \sum_{\chi \in \hat{A}} \sum_{a \in A} \sum_{b \in A} f(a)\overline{g(b)}\chi(a)\overline{\chi(b)} \\
&= \frac{1}{|A|} \sum_{a,b \in A} f(a)\overline{g(b)} \sum_{\chi \in \hat{A}} \overline{\delta_a(\chi)}\delta_b(\chi) \\
&= \frac{1}{|A|} \sum_{a,b \in A} f(a)\overline{g(b)} \, \langle \delta_b, \delta_a \rangle \\
&= \sum_{a \in A} f(a)\overline{g(a)} \\
&= \langle f, g \rangle,
\end{aligned}
$$

where we have applied Lemma 5.2.1 to the group $\hat{A}$. Next

$$
\begin{aligned}
\hat{\hat{f}}(\delta_a) &= \frac{1}{\sqrt{|A|}} \sum_{\chi \in \hat{A}} \hat{f}(\chi)\overline{\delta_a(\chi)} \\
&= \frac{1}{|A|} \sum_{\chi \in \hat{A}} \sum_{b \in A} f(b)\overline{\chi(b)}\overline{\chi(a)} \\
&= \frac{1}{|A|} \sum_{\chi \in \hat{A}} \sum_{b \in A} f(b^{-1})\chi(b)\overline{\chi(a)} \\
&= \frac{1}{|A|} \sum_{b \in A} f(b^{-1}) \, \langle \delta_b, \delta_a \rangle \\
&= f(a^{-1}).
\end{aligned}
$$

$\square$

5.3 Convolution

For functions on a finite abelian group there is a convolution product that resembles the convolution on the reals. Let f and g be in $\ell^2(A)$; we define their *convolution product* by

$$
f * g(a) = \frac{1}{\sqrt{|A|}} \sum_{b \in A} f(b)g(b^{-1}a).
$$

Theorem 5.3.1 *For $f, g \in \ell^2(A)$ we have*

$$\widehat{f * g} = \hat{f}\hat{g},$$

where on the right-hand side we have the pointwise product.

In particular, it follows that the convolution product is associative, distributive, and commutative; i.e.,

$$(f * g) * h = f * (g * h), \quad f * (g + h) = f * g + f * h, \quad f * g = g * f$$

holds for $f, g, h \in \ell^2(A)$.

Proof: We have

$$\widehat{f * g}(\chi) = \frac{1}{\sqrt{|A|}} \sum_{b \in A} f * g(b)\overline{\chi(b)} = \frac{1}{|A|} \sum_{a \in A} \sum_{b \in A} f(a)g(a^{-1}b)\overline{\chi(b)}.$$

Replacing b by ab gives

$$\widehat{f * g}(\chi) = \frac{1}{|A|} \sum_{a \in A} f(a)\overline{\chi(a)} \sum_{b \in A} g(b)\overline{\chi(b)} = \hat{f}(\chi)\hat{g}(\chi).$$

$\square$

5.4   Exercises

Exercise 5.1 Let A be a finite abelian group. Show that the Fourier transform of a character χ equals

$$\hat{\chi}(\eta) = \begin{cases} \sqrt{|A|} & \text{if } \eta = \chi, \\ 0 & \text{otherwise.} \end{cases}$$

Exercise 5.2 Show that for A and B finite abelian groups we have $\widehat{A \times B}$ isomorphic to $\hat{A} \times \hat{B}$. Conclude that for every finite abelian group A we have $|A| = |\hat{A}|$.

Exercise 5.3 Let A, B be finite abelian groups and let $\psi : A \to B$ be a group homomorphism. Show that the prescription

$$\psi^*(\chi) = \chi \circ \psi$$

defines a group homomorphism $\psi^* : \hat{B} \to \hat{A}$.

Exercise 5.4 Let A be a finite abelian group and B a subgroup. The restriction of characters gives a homomorphism res : $\hat{A} \to \hat{B}$, $\chi \mapsto \chi|_B$. Show that the kernel of res is isomorphic to the dual group $\widehat{A/B}$ of the quotient A/B, and conclude that res is surjective.

Exercise 5.5 Give an example of a finite abelian group A and a subgroup B such that there is no group C with $A \cong B \times C$.

Exercise 5.6 Let $1 \to A \to B \to C \to 1$ be an exact sequence of finite abelian groups. Use Exercises 5.3 and 5.4 to show that it induces an exact sequence

$$1 \to \hat{C} \to \hat{B} \to \hat{A} \to 1.$$

Exercise 5.7 Let $\chi_1, \chi_2, \ldots, \chi_n$ be distinct characters of the finite abelian group A. Show that $\chi_1, \chi_2, \ldots, \chi_n$ are linearly independent in the complex vector space of all maps from A to $\mathbb{C}$.

Chapter 6

LCA Groups

In this chapter we are going to set out the basic terminology of abstract harmonic analysis. This will require a modest amount of topology, which is introduced in the first section below.

6.1 Metric Spaces and Topology

For the reader's convenience we will briefly recall the basic properties of metrics, and the notions of continuity and topology.

Let X be a set. Recall that a *metric* on X is a map

$$d : X \times X \to [0, \infty)$$

such that

- d is *definite*, i.e.,

$$d(x, y) = 0 \iff x = y$$

  holds for every $x, y \in X$;

- d is *symmetric*, i.e.,

$$d(x, y) = d(y, x)$$

  for every $x, y \in X$; and

- d satisfies the *triangle inequality*, i.e.,

$$d(x, y) \leq d(x, z) + d(z, y)$$

for all $x, y, z \in X$.

A set X together with a metric d is called a *metric space*.

Examples.

- $X = \mathbb{R}$ with $d(x, y) = |x - y|$ is a metric space.

- Let $X = H$ be a Hilbert space; then $d(x, y) = \|x - y\|$ gives a metric on X (see Exercise 6.1).

- Let $X = \mathbb{R}/\mathbb{Z}$; then $d(x, y) = |e^{2\pi i x} - e^{2\pi i y}|$ defines a metric on X.

- On any set X we can establish the *discrete metric* as follows. Set

$$d(x, y) \;=\; \begin{cases} 0 & \text{if } x = y, \\ 1 & \text{if } x \neq y. \end{cases}$$

Lemma 6.1.1 *For three points x, y, z in a metric space X we have*

$$|d(x, y) - d(x, z)| \;\leq\; d(y, z).$$

Proof: Since $d(x, y) \leq d(x, z) + d(y, z)$ we get $d(x, y) - d(x, z) \leq d(y, z)$. Changing the roles of y and z gives $d(x, z) - d(x, y) \leq d(y, z)$. Together this implies the claim. $\qquad\qquad\square$

For a metric d, the geometric meaning of $d(x, y)$ is that of a distance between the points x and y. Let (X, d) be a metric space. A sequence (x_n) in X is said to *converge* to $x \in X$ if the sequence of distances $d(x_n, x)$ tends to zero. In other words, x_n tends to x if for every $\varepsilon > 0$ there is a natural number n_0 such that for all $n \geq n_0$,

$$d(x_n, x) \;<\; \varepsilon.$$

If x_n tends to x, then x is uniquely determined by the sequence x_n (Exercise 6.2), so it is justified to write

$$\lim_{n \to \infty} x_n \;=\; x.$$

Whenever the notation $\lim_{n\to\infty} x_n$ is used, it is implicitly stated that the limit exists, i.e., that the sequence (x_n) indeed converges.

Let X and Y be metric spaces. A map $f : X \to Y$ is called *continuous* if f maps convergent sequences to convergent sequences and preserves their limits, i.e., if

$$\lim_{n\to\infty} f(x_n) \;=\; f\left(\lim_{n\to\infty} x_n\right)$$

for every convergent sequence x_n in X. For functions $f : \mathbb{R} \to \mathbb{R}$ this notion coincides with the notion of continuity from analysis.

Examples.

- If X is discrete, then every map $f : X \to Y$ is continuous (Exercise 6.4).

- The natural projection $\mathbb{R} \to \mathbb{R}/\mathbb{Z}$ is continuous.

Let X be a set; then two metrics d_1 and d_2 on X are called *equivalent* if they define the same set of convergent sequences, i.e., $d_1 \sim d_2$ if for all sequences (x_n),

$$(x_n) \text{ converges in } d_1 \;\;\Leftrightarrow\;\; (x_n) \text{ converges in } d_2.$$

The following proposition describes an instance where this happens.

Proposition 6.1.2 *Let (X, d) be a metric space. Let $f : X \to X$ be a homeomorphism; i.e., f is continuous, bijective, and the inverse f^{-1} also is continuous. Then the metric*

$$d'(x, y) \;=\; d(f(x), f(y))$$

is equivalent to d.

Proof: The proof of the fact that d' is a metric is left to the reader (Exercise 6.6). Assume, then, that x_n is a sequence that converges in the metric d. Then $f(x_n)$ converges in d, since f is continuous, which means that x_n converges in d'. The inverse direction follows similarly by the fact that f^{-1} is continuous. $\qquad\square$

Examples.

- The metric

$$d(x, y) \;=\; \left| x^3 - y^3 \right|$$

  is equivalent to the standard metric on $\mathbb{R}$.

- The discrete metric on $\mathbb{R}$ is not equivalent to the standard metric on $\mathbb{R}$ (see Exercise 6.9).

Let (X, d) be a metric space. The *diameter* of (X, d) is defined to be

$$\mathrm{diam}(X) \;\overset{\mathrm{def}}{=}\; \sup_{x, y \in X} d(x, y).$$

The diameter can be a real number ≥ 0 or infinity. The next lemma shows that every metric is equivalent to one of finite diameter.

Lemma 6.1.3 *Let X be a set. For every metric d on X there is an equivalent metric d' with values in $[0, 1]$.*

Proof: Let d be a metric on X. We claim that

$$d'(x, y) \;=\; \frac{d(x, y)}{d(x, y) + 1}$$

is an equivalent metric. First we have to show that it is a metric at all. The map $f(x) = x/(x + 1)$ is a monotonic homeomorphism of $[0, \infty)$ to $[0, 1)$. Since $f(x) = 0$ is equivalent to $x = 0$, the positive definiteness of d' is clear. To see that the triangle inequality holds, let $a, b, c \geq 0$ satisfy $a \leq b + c$. We now have to show that then $f(a) \leq f(b) + f(c)$. If $a \leq b$, then $f(a) \leq f(b) \leq f(b) + f(c)$, so the claim follows. The same holds if $a \leq c$. So assume that $a \geq b, c$. Then $a \leq b + c$ implies

$$\frac{a}{a + 1} \;\leq\; \frac{b}{a + 1} + \frac{c}{a + 1} \;\leq\; \frac{b}{b + 1} + \frac{c}{c + 1}.$$

Since f is a homeomorphism, it follows that a sequence a_n in $[0, \infty)$ tends to zero if and only if $f(a_n)$ tends to zero. This implies that d and d' are equivalent. $\square$

A set X together with an equivalence class of metrics $[d]$ is called a *metrizable space* or a *metrizable topological space*.

Let $(X, [d])$ be a metrizable space. A *dense subset* of X is a subset $D \subset X$ such that for every $x \in X$ there is a sequence $y_n \in D$ such that y_n converges to x. The standard example of this is the subset $\mathbb{Q}$ of $\mathbb{R}$, which is a dense subset, since every real number can be approximated by rationals. If the dense subset is countable, we can choose a sequence y_n in D such that for every point $x \in X$ there is a subsequence of y_n that converges to x. Such a sequence is called a *dense sequence*.

Lemma 6.1.4 *Let X, Y be metrizable spaces. Let f, g be continuous maps from X to Y. If f and g agree on a dense subset D of X, then they are equal.*

Proof: Let $x \in X$; then there is a sequence $d_n \in D$ with limit x. Therefore, since f and g are continuous,

$$f(x) = f(\lim_n d_n) = \lim_n f(d_n) = \lim_n g(d_n) = g(\lim_n d_n) = g(x).$$

$\square$

For the purposes of this book the notion of a metrizable space is quite satisfactory, but for the convenience of the reader who may consult other sources we are obliged to give the connection of this concept to the more general notion of a topological space.

Let (X, d) be a metric space. A subset $U \subset X$ is called *open* if for every $u \in U$ there is a positive real number $r > 0$ such that the *open ball of radius r around u*,

$$B_r(u) = B_r^d(u) = \{x \in X \mid d(x, u) < r\},$$

is fully contained in U. The triangle inequality ensures that every open ball $B_r(u)$ is itself an open set, and thus the open sets are precisely the unions of open balls.

Let $\mathcal{O}$ be the set of all open sets in X, so $\mathcal{O}$ is a subset of the power set $\mathcal{P}(X)$. The following properties are immediate:

(a) $\emptyset, X \in \mathcal{O}$,

(b) $U, V \in \mathcal{O} \implies U \cap V \in \mathcal{O}$, and

(c) if $U_j \in \mathcal{O}$ for all j in some index set J, then the union $\bigcup_{j \in J} U_j$ also lies in $\mathcal{O}$.

A subset $\mathcal{O}$ of $\mathcal{P}(X)$ satisfying (a), (b), and (c) is called a *topology* on X. So we see that a metric d gives rise to a topology on X.

Lemma 6.1.5 *Two metrics d_1 and d_2 on a set X are equivalent if and only if they define the same topology on X.*

Proof: Let d_1 and d_2 be equivalent. We have to show that they define the same topology. Therefore, let $U \subset X$ be open with respect to d_1 and let $u \in U$.

Assume that there is no $r > 0$ such that the d_2-ball of radius r is contained in U. Then for every $n \in \mathbb{N}$ there is $x_n \in X \smallsetminus U$ such that $d_2(x_n, u) < 1/n$, which means that the sequence x_n converges in d_2 to u. Since $d_1 \sim d_2$ it also converges in d_1. Since U was open in d_1, there is $r > 0$ such that $d_1(x, u) < r \Rightarrow x \in U$ for every $x \in X$. Since x_n tends to u in d_1, there is $n \in \mathbb{N}$ such that $d_1(x_n, u) < r$, which implies $x_n \in U$, a contradiction!

Hence the assumption must be false, so U indeed contains some open d_2-ball around u, and since u was arbitrary, this implies that U is open with respect to d_2. By obvious symmetry the similar argument from d_2 to d_1 is also clear. Thus we have established that if $d_1 \sim d_2$, then the topologies agree.

For the inverse direction assume that d_1 and d_2 define the same open sets. Let x_n be a sequence that converges to x in d_1. The definition of convergence can be read as follows: For every $r > 0$ there is n_0 such that for $n \geq n_0$ we have $x_n \in B_r^{d_1}(x)$. Thus for every open set U containing x there is n_0 such that for $n \geq n_0$ we have $x_n \in U$. The d_2-balls of arbitrary radius $r > 0$ are open. Hence for every $r > 0$ there is n_0 such that for $n \geq n_0$ we have $x_n \in B_r^{d_2}(x)$, which implies that x_n converges to x in d_2. Again the direction from d_2-convergence to d_1-convergence follows by symmetry of the argument. $\qquad \square$

A set X together with a topology $\mathcal{O}$ will be called a *topological space*. The subsets $U \subset X$ that appear in the topology $\mathcal{O}$ are called *open sets*. The space $(X, \mathcal{O})$ is called *metrizable* if there is a metric d on X defining $\mathcal{O}$. So a metrizable space is either a set with a class of metrics or with a topology that is induced by a metric.[1]

---

[1]There are many topological spaces whaich are not metrizable. We only give one example: Let X be an infinite set and let $\mathcal{O}$ be the system of subsets U of X such that either $U = \emptyset$ or the complement $X \smallsetminus U$ is finite. Then $\mathcal{O}$ is a topology which is not derived from a metric.

Let X be a topological space and x a point in X. An *open neighborhood* of x is an open set $U \subset X$ that contains x. A *neighborhood* of x is a set $U \subset X$ that contains an open neighborhood of x.

Examples. The open interval $(-1, 1)$ is an open neighborhood of zero in $\mathbb{R}$. The intervals $[-1, 1)$, $(-1, 1]$, and $[-1, 1]$ are neighborhoods of zero in $\mathbb{R}$.

A subset $A \subset X$ of a topological space is called *closed* if its complement $X \smallsetminus A$ is open.

Lemma 6.1.6 *A subset A of a metrizable space X is closed if and only if for every sequence a_n in A that converges in X, the limit also lies in A.*

Proof: Let A be closed in X and let a_n be a convergent sequence lying in A. Assume that the limit x of the sequence does not lie in A; then it lies in the open set $U = X \smallsetminus A$. Then all but finitely many of the a_n must lie in the open set U, which is a contradiction. Hence the assumption is false, and thus $x \in A$.

For the other direction let $A \subset X$ be such that every sequence in A that converges in X already converges in A. Let $B = X \smallsetminus A$ be the complement of A. We have to show that B is open. So let $b \in B$ and *assume* that there is no ball $B_r(b)$ fully contained in B. Then, for every $n \in \mathbb{N}$ there is $x_n \in A = X \smallsetminus B$ such that $x_n \in B_{1/n}(b)$. This implies that the sequence x_n converges to b, and thus b must lie in A, a contradiction. So the assumption must be false, i.e., B indeed contains an open ball around b. Since b was arbitrary, B in fact is open, so A is closed. $\qquad\square$

Let $A \subset X$ be an arbitrary subset of the metrizable space X. The *closure* $\bar{A}$ of A is by definition the set of all limits of sequences in A that converge in X. It follows that $\bar{A}$ is the smallest closed subset containing A (see Exercise 6.17).

Examples. The closure of the interval $(0, 1)$ in $\mathbb{R}$ is the interval $[0, 1]$. The closure of the set $\mathbb{Q}$ in $\mathbb{R}$ is $\mathbb{R}$.

When $A \subset X$ is an arbitrary subset of a metrizable space, then every metric on X can be restricted to a metric on A, so A naturally becomes a metrizable space. Thus all notions connected with metrizable spaces can be applied to arbitrary subsets as well.

A metrizable space $(X, [d])$ is called *compact* if every sequence x_n in X has a convergent subsequence.[2]

Examples. Let $a < b$ be real numbers; then the interval $[a, b]$ is compact since we know from analysis that every sequence in this interval has a convergent subsequence. More generally, bounded closed subsets in $\mathbb{R}^n$ are compact (Exercise 6.11). A discrete space is compact if and only if it is finite (see Exercise 6.10).

Lemma 6.1.7 *Continuous images of compact sets are compact sets. In other words, if X and Y are compact metrizable spaces and the map $f : X \to Y$ is continuous, then the image $f(X)$ is a compact subset of Y.*

Proof: Let y_n be a sequence in $f(X)$. For each n choose a preimage $x_n \in X$ of y_n, i.e., an x_n such that $f(x_n) = y_n$. The sequence x_n then has a convergent subsequence x_{n_k}. Since f is continuous, it follows that $y_{n_k} = f(x_{n_k})$ is a convergent subsequence of y_n. $\square$

A metrizable space X is called σ-*compact* if there is a sequence $K_n \subset K_{n+1}$ of compact subsets such that $X = \bigcup_n K_n$. Such a sequence is called a *compact exhaustion* of X.

Examples.

- The real line is σ-compact, since $\mathbb{R}$ is the union of the compact intervals $[-n, n]$, $n \in \mathbb{N}$.

- Let X be a discrete space that is countable. Then X is σ-compact, since there are finite sets $K_n \subset K_{n+1}$ such that $X = \bigcup_n K_n$.

Finally, a metrizable space X is called *locally compact* if every point $x \in X$ has a compact neighborhood. This can be rephrased as follows: X is locally compact if given a metric d for X, for every point $x \in X$ there is an $r > 0$ such that the closed ball

$$\bar{B}_r(x) = \{y \in X | d(x, y) \le r\}$$

_____

[2]For a general topological space there is a different notion of compactness, which is discussed in Exercise 6.18.

is compact. Examples are $\mathbb{R}^n$ and discrete spaces. An example of a space that fails to be locally compact is an infinite-dimensional Hilbert space (see Exercise 6.21).

A space that is locally compact and σ-compact is also called *σ-locally compact*.

Examples. The spaces $\mathbb{R}$ and $\mathbb{R}/\mathbb{Z}$ are σ-locally compact, as is any countable discrete space.

6.2 Completion

In this section we will see that a general metric space can be completed in the same way as a separable pre-Hilbert space. As a consequence all pre-Hilbert spaces can be completed.

A *Cauchy sequence* in a metric space (X, d) is a sequence $(x_n)_{n \in \mathbb{N}}$ such that for every $\varepsilon > 0$ there is $n_0 \in \mathbb{N}$ with

$$m, n \geq n_0 \quad \Rightarrow \quad d(x_n, x_m) < \varepsilon.$$

Example. The sequence $x_n = \frac{1}{n}$ is a Cauchy sequence in $\mathbb{R}$. To see this let $\varepsilon > 0$ and choose $n_0 \in \mathbb{N}$ with $n_0 > 3/\varepsilon$. Then for $n, m \geq n_0$,

$$|x_n - x_m| = |\frac{1}{n} - \frac{1}{m}| \leq \frac{1}{n} + \frac{1}{m} \leq \frac{2}{n_0} < \varepsilon.$$

Lemma 6.2.1 *Every convergent sequence is a Cauchy sequence.*

Proof: Let (x_n) be a sequence in X, convergent to $x \in X$. Let $\varepsilon > 0$ and choose $n_0 \in \mathbb{N}$ such that for every $n \geq n_0$,

$$d(x_n, x) < \frac{\varepsilon}{2}.$$

Then, for any two $m, n \geq n_0$ we have

$$\begin{aligned} d(x_n, x_m) &\leq d(x_n, x) + d(x_m, x) \\ &< \frac{\varepsilon}{2} + \frac{\varepsilon}{2} = \varepsilon. \end{aligned}$$

Hence (x_n) is a Cauchy sequence. $\qquad \square$

The point about the definition of a Cauchy sequence becomes clear in the example $x_n = \frac{1}{n}$. This sequence already is a Cauchy sequence in the subspace $Y = \mathbb{R} \smallsetminus \{0\}$ but it does not converge in this subspace, since Y does not contain its limit zero. So Cauchy sequences detect "holes" in metric spaces. If such holes exist like in the case $\mathbb{R} \smallsetminus \{0\}$, they can be filled by plugging in new elements like zero in the example. So, completion means the construction of a new metric space $\bar{X}$ and an embedding $X \hookrightarrow \bar{X}$ in a way that $\bar{X}$ has no holes and the new points $\bar{X} \smallsetminus X$ fill the holes of X. For this we first need to make clear what we mean by embedding X into another space.

Let X, Y be metric spaces. An *isometry* from X to Y is a map $f \colon X \to Y$ with
$$d(x, x') = d(f(x), f(x'))$$
for any two elements $x, x' \in X$. An isometry is continuous. A map $g \colon X \to Y$ is called and *isometric isomorphism* if g is an isometry and g is bijective. In that case the inverse map g^{-1} also is an isometry.

A metric space X is called *complete* if every Cauchy sequence in X converges.

Theorem 6.2.2 *(Completion)*
Let X be a metric space. Then there exists an isometry $\varphi \colon X \to \bar{X}$ into a complete metric space $\bar{X}$ such that the image $\varphi(X)$ is dense in $\bar{X}$. The pair $(\bar{X}, \varphi)$ is called a completion *of X.*

The completion is uniquely determined in the following sense If $\psi \colon X \to Y$ is another isometry onto a dense subspace of a complete space Y, then there is a unique isometric isomorphism $\alpha \colon \bar{X} \to Y$ such that $\psi = \alpha \circ \varphi$. We illustrate this by the following commutative diagram:

Proof: Let (X, d) be a given metric space. We will first construct $\bar{X}$. Let $\tilde{X}$ be the set of all Cauchy sequences in X. We get a natural map $\tilde{\varphi} \colon X \to \tilde{X}$ mapping $x \in X$ to the constant sequence $x_n = x$.

On $\tilde{X}$ we introduce an equivalence relation as follows. We say that a sequence (x_n) is equivalent to (y_n) and we write this as $(x_n) \sim (y_n)$, if the sequence of numbers $d(x_n, y_n)$ tends to zero. Note that if (y_n) is a subsequence of $(x_n) \in \tilde{X}$, then $(y_n) \sim (x_n)$.

Now we define

$$\bar{X} \overset{\text{def}}{=} \tilde{X}/\sim,$$

so $\bar{X}$ is the set of equivalence classes in $\tilde{X}$.

We say that a Cauchy sequence (x_n) is a *strong Cauchy sequence* if

$$d(x_m, x_n) < \frac{1}{\min(m,n)}$$

holds for all $m, n \in \mathbb{N}$. Every Cauchy sequence has a subsequence which is strong.

Lemma 6.2.3 *Let (x_n) and (y_n) be in the space $\tilde{X}$, then the sequence $d(x_n, y_n)$ converges in $\mathbb{R}$ and its limit remains the same if (x_n) and (y_n) are replaced by equivalent sequences.*

Assume in particular that (x_n) and (y_n) are strong Cauchy sequences. Then for every $k \in \mathbb{N}$,

$$d(x_k, y_k) \leq \frac{2}{k} + \lim_n d(x_n, y_n)$$

Proof of the lemma. For $m, n \in \mathbb{N}$ we use Lemma 6.1.1 to estimate

$$|d(x_n, y_n) - d(x_m, y_m)|$$

$$
\begin{aligned}
&= \quad |d(x_n, y_n) - d(x_n, y_m) + d(x_n, y_m) - d(x_m, y_m)| \\
&\leq \quad |d(x_n, y_n) - d(x_n, y_m)| + |d(x_n, y_m) - d(x_m, y_m)| \\
&\leq \quad d(y_n, y_m) + d(x_n, x_m).
\end{aligned}
$$

So if (x_n) and $y_n)$ are Cauchy sequences it follows that $d(x_n, y_n)$ is a Cauchy sequence in $\mathbb{R}$ and since every Cauchy sequence in $\mathbb{R}$ converges, this sequence indeed converges. It is a simple consequence of the triangle inequality that each sequence can be replaced with an equivalent one. The last assertion follows from the estimate above.

□

We define a metric $\bar{d}$ on $\bar{X}$ as follows. Let $[x_n]$ denote the class of the Cauchy sequence (x_n) in $\bar{X}$. We define

$$\bar{d}([x_n], [y_n]) \overset{\text{def}}{=} \lim_{n \to \infty} d(x_n, y_n).$$

This limit exists and is unique by Lemma 6.2.3.

The proof that $\bar{d}$ indeed is a metric is perfectly straightforward. For instance, the triangle inequality is shown as follows,

$$
\begin{aligned}
\bar{d}([x_n],[y_n]) &= \lim_n d(x_n,y_n) \\
&\leq \lim_n d(x_n,z_n)+d(z_n,y_n) \\
&= \bar{d}([x_n],[z_n])+\bar{d}([z_n],[y_n]).
\end{aligned}
$$

We define $\varphi\colon X \to \bar{X}$ by

$$
\varphi(x) \overset{\text{def}}{=} [\tilde{\varphi}(x)].
$$

So we get $\varphi(x) = [x_n]$ with $x_n = x$ for every $n \in \mathbb{N}$. It follows

$$
\bar{d}(\varphi(x),\varphi(y)) = \lim_n d(x,y) = d(x,y),
$$

so φ is an isometry. To see that the image of φ is dense in $\bar{X}$ pick $[x_n] \in \bar{X}$ and let $\varepsilon > 0$. Choose $n_0 \in \mathbb{N}$ such that for $m,n \geq n_0$ we have $d(x_n,x_m) < \varepsilon/2$. Let $x = x_{n_0}$. Then

$$
\bar{d}(\varphi(x),[x_n]) = \lim_n d(x_{n_0},x_n) \leq \varepsilon/2 < \varepsilon.
$$

Since $\varepsilon > 0$ is arbitrary this implies that $\varphi(X)$ is dense in $\bar{X}$.

We next have to show that $\bar{X}$ is complete. For this let $([x^k])_{k\in\mathbb{N}} = ([x_n^k])_{k\in\mathbb{N}}$ be a Cauchy sequence in $\bar{X}$ indexed by k. That means that for each $k \in \mathbb{N}$ we have a Cauchy sequence $(x_n^k)_{n\in\mathbb{N}}$ in X. Replacing $(x_n^k)_{n\in\mathbb{N}}$ by a subsequence if necessary we can assume that $(x_n^k)_n$ is a strong Cauchy sequence. Further, replacing $([x^k])$ with a subsequence we may assume that it is a strong Cauchy sequence as well. Set $y_j = x_j^j$, by Lemma 6.2.3 we have

$$
\begin{aligned}
d(y_i,y_j) &= d(x_i^i,x_j^j) \\
&\leq d(x_i^i,x_i^j)+d(x_i^j,x_j^j) \\
&< \frac{2}{i}+\bar{d}([x^i],[x^j])+\frac{1}{\min(i,j)} \\
&< \frac{2}{i}+\frac{2}{\min(i,j)}
\end{aligned}
$$

So (y_j) is a Cauchy sequence in X. hence it defines an element $[y]$ of $\bar{X}$. We want to show that the sequence $[x^k]$ converges to $[y]$. This follows from

$$
\begin{aligned}
\bar{d}([x^k], [y]) &= \lim_j d(x_j^k, x_j^j) \\
&\leq \lim_j \frac{2}{j} + \frac{1}{\min(j, k)} \\
&= \frac{1}{k}.
\end{aligned}
$$

To finish the proof of Theorem 6.2.2 assume now that we have a second completion $\psi \colon X \to Y$ as in the theorem. For $\bar{X} \in \bar{X}$ choose a sequence $x_n \in X$ such that $\varphi(x_n)$ converges to $\bar{X}$. Define

$$
\alpha(\bar{X}) \stackrel{\text{def}}{=} \lim_n \psi(x_n).
$$

This needs explanation. First, since φ is isometric the sequence x_n is Cauchy and thus $\psi(x_n)$ is Cauchy, hence convergent, so the limit exists. It is easy to see that the limit does not depend on the choice of the sequence x_n. Hence α is well defined. Further, it is straightforward to see that that α is isometric. The inverse map α^{-1} is defined in the same way with reversed roles, so

$$
\alpha^{-1}(y) \stackrel{\text{def}}{=} \lim_n (\varphi(x_n)),
$$

where x_n is an arbitrary sequence in X with $\psi(x_n) \to y$. Then $\alpha \alpha^{-1} = \text{Id}$ and $\alpha^{-1}\alpha = \text{Id}$. By density arguments the map α is uniquely determined. $\qquad\square$

If the metric space X carries additional structure, this is often preserved in its completion $\bar{X}$. For example the completion of a pre-Hilbert space is a Hilbert space. Also consider $L_{bc}^1(\mathbb{R})$ with the metric $d(f, g) \stackrel{\text{def}}{=} \|f - g\|_1$. Then its completion

$$
L^1(\mathbb{R}) \stackrel{\text{def}}{=} \overline{L_{bc}^1(\mathbb{R})}
$$

preserves the structure of a vector space. For example if $f, g \in L^1(\mathbb{R})$ and (f_n) and (g_n) are sequences converging to f and g resp., then their sum can be defined by

$$f + g \overset{\text{def}}{=} \lim_n (f_n + g_n).$$

Also the norm $\|.\|_1$ extends to $L^1(\mathbb{R})$ and makes it a complete normed vector space.

6.3   LCA Groups

A *metrizable abelian group* is an abelian group A together with a class of metrics [d] (or a topology that comes from a metric) such that the multiplication and inversion,

$$\begin{array}{ccc} A \times A & \to & A, \\ (x, y) & \mapsto & xy, \end{array} \quad \text{and} \quad \begin{array}{ccc} A & \to & A, \\ x & \mapsto & x^{-1}, \end{array}$$

are continuous. In other words, we insist that when x_n is a sequence converging to x and y_n converges to y, then the sequence $x_n y_n$ converges to xy and the sequence x_n^{-1} converges to x^{-1}.

Examples.

- Any group with the discrete metric is a metrizable group.

- We write $\mathbb{R}^\times$ for the set $\mathbb{R} \smallsetminus \{0\}$. The groups $(\mathbb{R}, +)$ and $(\mathbb{R}^\times, *)$ with the topology of $\mathbb{R}$ are examples of metrizable groups, since if x_n, y_n are sequences of real numbers converging to x and y, respectively, then $x_n + y_n$ converges to $x + y$, and $-x_n$ converges to $-x$. A similar result holds for $\mathbb{R}^\times$.

- The group $\mathbb{R}/\mathbb{Z}$ with the metric given in the previous section is a metrizable group.

A metrizable σ-locally compact abelian group is called an *LCA group*.

Examples.

- Any countable abelian group with the discrete metric is an LCA group (see Exercise 6.15).

- The groups $\mathbb{R}$ and $\mathbb{R}/\mathbb{Z}$ are LCA groups.

Lemma 6.3.1 *An LCA group contains a countable dense subset.*

Proof: This result is a consequence of the σ-compactness. Let $A = \bigcup_{n \in \mathbb{N}} K_n$ be a compact exhaustion of A and choose a metric for A. By Exercise 6.18, K_1 can be covered by finitely many open balls of radius 1. Let $a_1, \ldots, a_{r_1}$ be their centers. Next, K_2 can be covered by finitely many open balls of radius $1/2$ with centers $a_{r_1+1}, \ldots, a_{r_2}$ and so on, so K_j can be covered by finitely many open balls of radius $1/j$ with centers $a_{r_{j-1}+1}, \ldots, a_{r_j}$. The sequence (a_k) thus constructed is dense in A. □

Let A be an LCA group. A compact exhaustion (K_n) of A is called *absorbing* if for every compact set $K \subset A$ there is an index $n \in \mathbb{N}$ such that $K \subset K_n$; i.e., the exhaustion absorbs all compact sets.

Examples.
The exhaustion $K_n = [-n, n]$ of $\mathbb{R}$ is absorbing, since every compact subset of $\mathbb{R}$ is bounded. The exhaustion $K_n = [-n, n] \smallsetminus (0, \frac{1}{n})$ is not absorbing, since no K_n contains the compact interval $[0, 1]$.

Lemma 6.3.2 *Let A be an LCA group; then there exists an absorbing exhaustion.*

Proof: Let A be an LCA group and let U be an open neighborhood of the unit such that its closure $V = \bar{U}$ is compact. Let L_n be a given compact exhaustion. Then for each n, the set $K_n = VL_n = \{vl | v \in V, l \in L_n\}$ is compact again, since it is the image of the compact set $V \times L_n$ under the multiplication map, which is continuous. Since $L_n \subset K_n$ we infer that the sequence (K_n) again forms a compact exhaustion. To show that it is absorbing let $K \subset A$ be a compact subset.

Assume that for each $n \in \mathbb{N}$ there is $x_n \in K$ that is not in K_n. Since K is compact the sequence x_n has a convergent subsequence. We may replace x_n by this subsequence to assume that $x_n \to x$. Since (L_n) is an exhaustion, there is $n_0 \in \mathbb{N}$ such that $x \in L_{n_0}$. The set Ux is an open neighborhood of x, so there exists n_1 such that $x_n \in Ux$ for $n \geq n_1$. For $n \geq n_0, n_1$ we infer

$$x_n \in Ux \subset UL_n \subset VL_n = K_n,$$

which is a contradiction. Hence the assumption is wrong, so $K \subset K_n$ for some n, i.e., the exhaustion K_n is absorbing. □

6.4   Exercises

Exercise 6.1 Let V be a vector space with a norm $\|.\|$. Show that $d(x,y) = \|x - y\|$ defines a metric on V.

Exercise 6.2 Let x_n be a sequence in the metric space X that converges to $x \in X$ and to $y \in X$. Show that $x = y$.

Exercise 6.3 Let X be a discrete metric space, i.e., the metric is the discrete metric. Show that a sequence (x_n) in X converges to x if and only if it becomes stationary, i.e., if there is $n_0 \in \mathbb{N}$ such that for all $n \geq n_0$ we have $x_n = x$.

Exercise 6.4 Let X be a discrete space. Let Y be a metric space and $f : X \to Y$ an arbitrary map. Show that f is continuous.

Exercise 6.5 Show that every finite subgroup of $\mathbb{R}/\mathbb{Z}$ is cyclic.

Exercise 6.6 Let (X,d) be a metric space. Let $f : X \to X$ be an injective map. Show that

$$d'(x,y) = d(f(x), f(y))$$

defines a metric on X.

Exercise 6.7 Let X, Y, Z be metric spaces and let $g : X \to Y$ and $f : Y \to Z$ be continuous maps. Show that the composition $f \circ g : X \to Z$ is continuous.

Exercise 6.8 Let X be a metrizable space. Show that the following are equivalent:

(a) X is locally compact.

(b) Every $x \in X$ has an open neighborhood U such that the closure $\bar{U}$ is compact.

(c) Every $x \in X$ has an open neighborhood U such that there is a compact subset C of X that contains U.

Exercise 6.9 Show that the discrete metric is not equivalent to the standard metric on $\mathbb{R}$.

Exercise 6.10 Show that a discrete space is compact if and only if it is finite.

Exercise 6.11 On the real vector space $\mathbb{R}^n$, for a natural number n, we define the Euclidean norm by

$$\|a\|_2 = \sqrt{a_1^2 + a_2^2 + \cdots + a_n^2}, \quad a \in \mathbb{R}^n.$$

Show that

(a) $d(a,b) = \|a - b\|_2$ defines a metric on $\mathbb{R}^n$,

(b) a sequence $(a^{(j)})_{j \in \mathbb{N}}$ in $\mathbb{R}^n$ converges if and only if every entry $a_k^{(j)}$ for $k = 1, \ldots n$ converges,

(c) a subset $A \subset \mathbb{R}^n$ is compact if and only if A is closed and bounded, i.e., there is a $C > 0$ such that $\|a\|_2 \leq C$ for every $a \in A$.

Exercise 6.12 Let X be an infinite set. Define a subset $\mathcal{O}$ of the power set $\mathcal{P}(X)$ by

$$A \in \mathcal{O} \Leftrightarrow A = \emptyset \text{ or } X \smallsetminus A \text{ is finite.}$$

Show that $\mathcal{O}$ is a topology.

Exercise 6.13 Let (x_n) be a Cauchy sequence in a metric space (X, d). Let $(x_{n_k})_{k \in \mathbb{N}}$ be a subsequence. Show that (x_n) converges if and only if (x_{n_k}) converges. and in that case the limits agree.

Exercise 6.14 Let X be a metric space and $\varphi \colon X \to \bar{X}$ its completion. Show that for every complete metric space Y and every isometry/continuous map $\psi \colon X \to Y$ there exists a unique isometry/continuous map $\alpha \colon \bar{X} \to Y$ such that $\psi = \alpha \circ \varphi$.

Exercise 6.15 Show that every countable discrete group is an LCA group.

Exercise 6.16 For $f, g \in \mathcal{S}(\mathbb{R})$ let

$$d(f,g) = \sum_{m,n=0}^{\infty} \frac{1}{2^{m+n}} \frac{\sigma_{m,n}(f-g)}{1 + \sigma_{m,n}(f-g)}.$$

Show that d is a metric on the space $\mathcal{S}(\mathbb{R})$ and that a sequence (f_j) converges to $f \in \mathcal{S}$ in this metric iff for every two $m, n \geq 0$ the sequence $\sigma_{m,n}(f - f_j)$ tends to zero as $j \to \infty$.

Exercise 6.17 Show that the closure of a subset $A \subset X$ of a metric space is the smallest closed subset containing A; i.e., show that $\bar{A}$ is closed and that each closed set B that contains A also contains $\bar{A}$.

Exercise 6.18 Let X be a topological space. An *open covering* of X is a family $(U_j)_{j \in J}$ of open sets such that $X = \bigcup_{j \in J} U_j$. A *subcovering* is a subfamily $(U_j)_{j \in F}$ that still is a covering; i.e., $X = \bigcup_{j \in F} U_j$, where F is a subset of the index set J. A subcovering is called a *finite subcovering*, if F is a finite set. Now X is called a *compact space* if every open covering of X admits a finite subcovering. Show that for a metrizable space X this definition of compactness coincides with the one given in the text; i.e., show that every sequence in X has a convergent subsequence if and only if every open covering has a finite open subcovering.

Exercise 6.19 For $j \in \mathbb{N}$ let A_j be a nontrivial compact abelian group. Let

$$A = \prod_{j \in \mathbb{N}} A_j.$$

Let d_j be a metric on A_j such that A_j has diameter 1 and define

$$d(x, y) = \sum_{j \in \mathbb{N}} \frac{1}{2^j} d_j(x_j, y_j).$$

(a)   Show that $d(x, y)$ defines a metric on A that makes A a compact LCA group.

(b)   Show that for each $j \in \mathbb{N}$ the projection $A \to A_j$ is a continuous group homomorphism.

(c)   Show that if each A_j is finite, then every continuous group homomorphism $\mathbb{R} \to A$ is trivial.

Exercise 6.20 For $j \in \mathbb{N}$ let A_j be a compact group. Suppose that for $i < j$ there is a continuous group homomorphism $\varphi_i^j : A_j \to A_i$. Suppose that $\varphi_i^k \circ \varphi_k^j = \varphi_i^j$ whenever $i < k < j$. Let $\varprojlim A_j$ be the set of all $x \in \prod_j A_j$ such that $x_i = \varphi_i^j(x_j)$ for all $i < j$.

(a) Show that $\varprojlim A_j$ is a closed subgroup of $\prod_j A_j$. This group is called the *projective limit* of the A_j.

(b) Show that the projections induce continuous group homomorphisms

$$p_i : \varprojlim A_j \to A_i$$

for $i \in \mathbb{N}$ that satisfy $\varphi_i^k \circ p_k = p_i$ whenever $k > i$.

(c) Suppose there is a compact abelian group A and a sequence of continuous group homomorphisms $q_i : A \to A_i$ such that $\varphi_i^k \circ q_k = q_i$ whenever $k > i$. Show that there is a unique continuous group homomorphism $\alpha : A \to \varprojlim A_j$ such that for each i the diagram

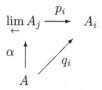

is commutative. This property is called the universal property of the projective limit.

Exercise 6.21 Show that an infinite-dimensional separable Hilbert space fails to be locally compact.

(Hint: Let $(e_n)_{n \in \mathbb{N}}$ be an orthonormal system. Show that no subsequence of (e_n) can be convergent.)

Exercise 6.22 Show that every open subgroup of a metrizable group is also closed.

(Hint: Let H be an open subgroup of G. Write G as a union of left cosets of H.)

Exercise 6.23 For $j \in \mathbb{N}$ let B_j be a discrete group. Suppose that for $i < j$ there is a group homomorphism $\psi_i^j : B_i \to B_j$ such that $\psi_j^k \circ \psi_k^i = \psi_j^i$ whenever $i < k < j$. For $b \in B_j$ and $k \geq j$ let $b_k = \psi_j^k(b)$, so b defines a sequence $(b_k)_{k \geq j}$. Indeed, the union $\cup_j B_j$ can be identified with the set of all these sequences. For $b \in B_j$ and $b' \in B_{j'}$ define $b \sim b'$ if and only if there is a $k \geq j, j'$ with $b_k = b'_k$.

(a) Show that $\sim$ is an equivalence relation on $\cup_j B_j$. Show that the quotient

$$\lim_{\to} B_j \; = \; \bigcup_j B_j / \sim$$

becomes an abelian group with the multiplication $[(b_k)][(b'_k)] = [(b_k b'_k)]$. This group is called the direct limit of the B_j.

(b) Show that the maps $e_i : B_i \to \lim_{\to} B_j$ given by $e_i(b) = [(b_k)]$ are group homomorphisms satisfying $e_k \circ \psi_i^k = e_i$ whenever $k > i$.

(c) Suppose there is a discrete abelian group B and a sequence of group homomorphisms $f_i : B_i \to B$ such that $f_k \circ \psi_i^k = f_i$ whenever $k > i$. Show that there is a unique group homomorphism $\beta : \lim_{\to} B_j \to B$ such that for each i the diagram

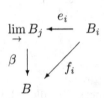

is commutative. This property is called the universal property of the direct limit.

Chapter 7

The Dual Group

In this chapter we further develop the general theory of LCA groups in showing that the group of all characters of a given LCA group is again an LCA group in a natural way. This then paves the way for Pontryagin duality.

7.1 The Dual as LCA Group

Let A be an LCA group. Recall that this means that A is a σ-compact, metrizable, locally compact abelian group.

A *character* of a metrizable abelian group A is a continuous group homomorphism $\chi : A \to \mathbb{T}$. The set of all characters of A is denoted by $\hat{A}$.

Proposition 7.1.1 *The characters of our standard examples are given as follows.*

(a) *The characters of the group $\mathbb{Z}$ are given by $k \mapsto e^{2\pi i k x}$, where $x \in \mathbb{R}/\mathbb{Z}$.*

(b) *The characters of $\mathbb{R}/\mathbb{Z}$ are given by $x \mapsto e^{2\pi i k x}$, where $k \in \mathbb{Z}$.*

(c) *The characters of $\mathbb{R}$ are given by $x \mapsto e^{2\pi i x y}$, where $y \in \mathbb{R}$.*

Proof: To prove (a), let $\varphi : \mathbb{Z} \to \mathbb{T}$ be a character. Then $\varphi(1) = e^{2\pi i x}$ for some $x \in \mathbb{R}/\mathbb{Z}$, and so for $k \in \mathbb{Z}$ arbitrary we get $\varphi(k) = \varphi(1)^k = e^{2\pi i k x}$.

101

For (c), let $\varphi : \mathbb{R} \to \mathbb{T}$ be a character. By continuity there is $\varepsilon > 0$ such that $\varphi([-\varepsilon, \varepsilon]) \subset \{\mathrm{Re}(z) > 0\}$. Let y be the unique element of $\left[-\frac{1}{4\varepsilon}, \frac{1}{4\varepsilon}\right]$ such that $\varphi(\varepsilon) = e^{2\pi i \varepsilon y}$. Then we claim that

$$\varphi\left(\frac{\varepsilon}{2}\right) = e^{2\pi i \frac{\varepsilon}{2} y}.$$

To prove the claim note that $\varphi(\frac{\varepsilon}{2})^2 = \varphi(\varepsilon) = e^{2\pi i \varepsilon y}$, so $\varphi(\frac{\varepsilon}{2}) = \pm e^{2\pi i \frac{\varepsilon}{2} y}$, and $-e^{2\pi i \frac{\varepsilon}{2} y}$ does not have positive real part.

Iterating this argument gives $\varphi\left(\frac{\varepsilon}{2^n}\right) = e^{2\pi i \frac{\varepsilon}{2^n} y}$, and so for $k \in \mathbb{Z}$ we get

$$\varphi\left(\frac{k}{2^n}\varepsilon\right) = \varphi\left(\frac{\varepsilon}{2^n}\right)^k = e^{2\pi i \frac{k}{2^n} \varepsilon y}.$$

The set of all rational numbers of the form $k/2^n$, $k \in \mathbb{Z}, n \in \mathbb{N}$, is dense in $\mathbb{R}$. Since φ is continuous, we conclude by Lemma 6.1.4 that $\varphi(x) = e^{2\pi i x y}$ for every $x \in \mathbb{R}$ as claimed. Finally, (b) follows from the fact that the characters of $\mathbb{R}/\mathbb{Z}$ are precisely the characters of $\mathbb{R}$ that send $\mathbb{Z}$ to 1. $\qquad\square$

Let A be an LCA group. Let $\hat{A}$ be the set of all characters of A.

Lemma 7.1.2 *The pointwise product $\chi\eta(a) = \chi(a)\eta(a)$ makes $\hat{A}$ an abelian group, called the dual group, or the Pontryagin dual, of A.*

Proof: Let $\chi, \eta \in \hat{A}$. As in the proof of Lemma 5.1.2 it can be shown that $\chi\eta$ and χ^{-1} are group homomorphisms. To see that $\chi\eta$ is a continuous map, let a_n be a sequence in A converging to $a \in A$. Then $\chi\eta(a_n) = \chi(a_n)\eta(a_n)$. Since $\chi(a_n)$ converges to $\chi(a)$, and likewise for η, it follows that $\chi\eta(a_n)$ converges to $\chi(a)\eta(a) = \chi\eta(a)$. Analogously one finds that χ^{-1} is continuous. This implies that $\hat{A}$ is an abelian group. $\qquad\square$

Examples.

- The dual group of $\mathbb{Z}$ is isomorphic to $\mathbb{R}/\mathbb{Z}$.

- The dual group of $\mathbb{R}/\mathbb{Z}$ is isomorphic to $\mathbb{Z}$.

- The dual group of $\mathbb{R}$ is isomorphic to $\mathbb{R}$.

To prove these statements one has to show that the bijections given in Proposition 7.1.1 are group homomorphisms. We treat only the first case, since the others are similar. So let φ be the map from $\mathbb{R}/\mathbb{Z}$ to $\hat{\mathbb{Z}}$ given by

$$\varphi(x)(k) = \varphi_x(k) = e^{2\pi i k x}.$$

For every $x, y \in \mathbb{R}/\mathbb{Z}$ we then have

$$\varphi_{x+y}(k) = e^{2\pi i k(x+y)} = e^{2\pi i k x}e^{2\pi i k y} = \varphi_x(k)\varphi_y(k),$$

which implies the claim.

We will now show that for a given LCA group A the dual $\hat{A}$ is an LCA group again.

Fix an absorbing compact exhaustion $A = \bigcup_{n \in \mathbb{N}} K_n$. For $\chi, \eta \in \hat{A}$ and $n \in \mathbb{N}$ set

$$\hat{d}_n(\chi, \eta) = \sup_{x \in K_n} |\chi(x) - \eta(x)|$$

and

$$\hat{d}(\chi, \eta) = \sum_{n=1}^{\infty} \frac{1}{2^n}\hat{d}_n(\chi, \eta).$$

Lemma 7.1.3 *The function $\hat{d}$ is a metric on the set $\hat{A}$.*

As we have constructed it, this metric is bounded by 2.

Proof: We will show only the triangle inequality because the other properties are obvious. For $\chi, \eta, \alpha \in \hat{A}$ we compute

$$
\begin{aligned}
\hat{d}_n(\chi, \eta) &= \sup_{x \in K_n} |\chi(x) - \eta(x)| \\
&= \sup_{x \in K_n} |\chi(x) - \alpha(x) + \alpha(x) - \eta(x)| \\
&\leq \sup_{x \in K_n} |\chi(x) - \alpha(x)| + \sup_{x \in K_n} |\alpha(x) - \eta(x)| \\
&= \hat{d}_n(\chi, \alpha) + \hat{d}_n(\alpha, \eta),
\end{aligned}
$$

and so

$$
\begin{aligned}
\hat{d}(\chi, \eta) &= \sum_{n=1}^{\infty} \frac{1}{2^n}\hat{d}_n(\chi, \eta) \leq \sum_{n=1}^{\infty} \frac{1}{2^n}\left(\hat{d}_n(\chi, \alpha) + \hat{d}_n(\alpha, \eta)\right) \\
&= \hat{d}(\chi, \alpha) + \hat{d}(\alpha, \eta).
\end{aligned}
$$

$\square$

Theorem 7.1.4 *With the metric above, the group $\hat{A}$ is a topological abelian group. A sequence χ_n converges in this metric if and only if it converges locally uniformly, so the metric class or topology does not depend on the exhaustion chosen. With this topology $\hat{A}$ is an LCA group.*

Proof: First we have to show that the group operations on $\hat{A}$ are continuous. For this let χ_j and η_j be sequences in $\hat{A}$ converging to χ and η, respectively. Then, for every natural number n,

$$
\begin{aligned}
\hat{d}_n(\chi_j\eta_j, \chi\eta) &= \sup_{x \in K_n} |\chi_j(x)\eta_j(x) - \chi(x)\eta(x)| \\
&= \sup_{x \in K_n} |(\chi_j(x) - \chi(x))\eta_j(x) + \chi(x)(\eta_j(x) - \eta(x))| \\
&\leq \sup_{x \in K_n} |\chi_j(x) - \chi(x)| + \sup_{x \in K_n} |\eta_j(x) - \eta(x)| \\
&= \hat{d}_n(\chi_j, \chi) + \hat{d}_n(\eta_j, \eta).
\end{aligned}
$$

Multiplying this by $1/2^n$ and summing over n gives

$$
\hat{d}(\chi_j\eta_j, \chi\eta) \leq \hat{d}(\chi_j, \chi) + \hat{d}(\eta_j, \eta).
$$

Thus $\chi_j\eta_j$ tends to $\chi\eta$, and it follows that multiplication is continuous. The inversion is dealt with in a similar fashion, so it follows that $(\hat{A}, [d])$ is a metrizable abelian group. Now, a sequence χ_j in $\hat{A}$ converges if and only if it converges uniformly on each K_n. Since the exhaustion (K_n) was absorbing, this means that the sequence converges if and only if it converges uniformly on every compact subset of A, which is equivalent to locally uniform convergence, since A is locally compact.

It remains to show that $\hat{A}$ is locally compact and σ-compact. Let d be a metric on A giving the fixed topology. For $r > 0$ let $B_r = B_r(1) = \{a \in A | d(a, 1) < r\}$. Set

$$
L_n = \{\chi \in \hat{A} \mid \chi(B_{\frac{1}{n}}) \subset \{\mathrm{Re}(z) \geq 0\}\}.
$$

We will show that the L_n form a compact exhaustion.

Lemma 7.1.5 *Let $n \in \mathbb{N}$. For every $\varepsilon > 0$ there is $\delta > 0$ such that for every $\chi \in L_n$,*

$$
\chi(B_\delta) \subset \{z \in \mathbb{T} : |z - 1| < \varepsilon\}.
$$

Proof: Let $n \in \mathbb{N}$ and $\varepsilon > 0$. For $k \in \mathbb{N}$ and $r > 0$ let

$$(B_r)^k = \{x_1 \cdot x_2 \cdots x_k \in A \mid x_1, x_2, \ldots, x_k \in B_r\}.$$

The space $A \times A$ is a metric space with the metric

$$d((a, a'), (b, b')) = d(a, b) + d(a', b').$$

The ε-δ criterion of continuity for the multiplication map implies that there is $\delta > 0$ such that

$$d((a, 1), (b, 1)) < 2\delta \implies d(ab, 1) < \frac{1}{n},$$

which implies that

$$(B_\delta)^2 \subset B_{\frac{1}{n}}.$$

Iteration of this argument gives that for each $k \in \mathbb{N}$ there is $\delta_k > 0$ such that

$$(B_{\delta_k})^k \subset B_{\frac{1}{n}}.$$

Let $k \in \mathbb{N}$ be so large that $|e^{\frac{\pi}{2k}i} - 1| < \varepsilon$. We claim that δ_k satisfies the condition of the lemma. For this let $x \in B_{\delta_k}$; then the elements $x, x^2, \ldots, x^k$ are all in $B_{\frac{1}{n}}$, and therefore for every $\chi \in L_n$ the real parts of $\chi(x), \chi(x)^2, \ldots, \chi(x)^k$ are all nonnegative. This can be so only if $\chi(x) = e^{\frac{\pi}{2k}it}$ for some $t \in [-1, 1]$, which implies $|\chi(x) - 1| < \varepsilon$. So we have shown that

$$\chi(B_{\delta_k}) \subset \{z \in \mathbb{T} : |z - 1| < \varepsilon\}.$$

$\square$

From this lemma we now deduce that L_n is compact. Let χ_j be a sequence in L_n. We have to find a convergent subsequence, that is, a locally uniformly convergent subsequence. Let a_j be a dense sequence in A as in Lemma 6.3.1. Since the sequence $(\chi_j(a_1))_j$ takes values in the compact set $\mathbb{T}$, there is a subsequence χ_j^1 of χ_j such that the sequence $\chi_j^1(a_1)$ converges. Next there is a subsequence χ_j^2 of χ_j^1 such that $\chi_j^2(a_2)$ converges. Iterating we find for each $k \in \mathbb{N}$ a subsequence χ_j^k of χ_j such that $\chi_j^k(a_1), \ldots, \chi_j^k(a_k)$ all converge. The diagonal sequence χ_j^j is a subsequence of χ_j, and all the sequences $\chi_j^j(a_k)$ for varying k converge. This defines a map $\chi : \{a_k \mid k \in \mathbb{N}\} \to \mathbb{T}$ given by

$$\chi(a_k) = \lim_{j \to \infty} \chi_j^j(a_k).$$

Now let $\varepsilon > 0$ and $\delta > 0$ as in Lemma 7.1.5. Assume $a_k^{-1}a_l \in B_\delta$. Then by Lemma 7.1.5 we have

$$|\chi_j^j(a_k) - \chi_j^j(a_l)| = |\chi_j^j(a_k^{-1}a_l) - 1| < \varepsilon,$$

and therefore

$$|\chi(a_k) - \chi(a_l)| \leq \varepsilon.$$

This implies that χ, being the locally uniform limit of the sequence χ_j^j, extends to a unique continuous map $\chi : A \to \mathbb{T}$. Since all the χ_j^j are group homomorphisms, then so is χ, which therefore is a character. Being the limit of the χ_j^j, the character χ still lies in L_n, and hence the latter is compact. The lemma and the theorem follow. □

Proposition 7.1.6 *The group isomorphisms $\mathbb{R}/\mathbb{Z} \to \hat{\mathbb{Z}}$, $\mathbb{Z} \to \widehat{\mathbb{R}/\mathbb{Z}}$, and $\mathbb{R} \to \hat{\mathbb{R}}$ given in Proposition 7.1.1 are homeomorphisms; i.e., they are continuous and so are their inverse maps. So in particular, we can say that $\hat{\mathbb{R}}$ is isomorphic to $\mathbb{R}$ as an LCA group.*

Proof: We will consider only the case of the isomorphism $\varphi : \mathbb{R} \to \hat{\mathbb{R}}$, $x \mapsto \varphi_x$, since the others are similar. To see that φ is continuous, let x_n be a sequence in $\mathbb{R}$, convergent to, say, $x \in \mathbb{R}$. Then, for every $y \in \mathbb{R}$, we have

$$|\varphi_{x_n}(y) - \varphi_x(y)| = |e^{2\pi i x_n y} - e^{2\pi i x y}|$$
$$= \left| \int_x^{x_n} 2\pi i y\, e^{2\pi i t y} dt \right| \leq 2\pi |y|\, |x_n - x|.$$

This implies that on every bounded interval the sequence of functions φ_{x_n} will converge uniformly to the function φ_x; hence we have that φ_{x_n} converges to φ_x locally uniformly on $\mathbb{R}$. We conclude that the map φ is continuous.

Next we prove that the inverse φ^{-1} is continuous. For this let x_n be a sequence in $\mathbb{R}$ such that φ_{x_n} is convergent in $\hat{\mathbb{R}}$ to, say, φ_x. We have to show that x_n converges to x in $\mathbb{R}$. Let $y \in \mathbb{R}$, $|y| \leq 1$. Then $\varphi_{x_n}(y) = e^{2\pi i x_n y}$ converges to $e^{2\pi i x y}$ uniformly in y. This implies that there are $k_n \in \mathbb{Z}$ such that $(x_n - x)y = k_n + \varepsilon_n$, where the sequence ε_n tends to zero in $\mathbb{R}$. Since this is true for every $y \neq 0$, the sequence x_n must be bounded. Hence there is a convergent subsequence x_{n_k}. Let x' be its limit. Then by the first part we know that $\varphi_{x_{n_k}}$ tends to $\varphi_{x'}$, which implies that $x' = x$. Since this holds for every convergent subsequence, it follows that x_n converges to x as claimed. □

7.2 Pontryagin Duality

In this section we state the result that every LCA group is canonically isomorphic to its bidual, i.e., the dual of the dual.

Proposition 7.2.1 *If A is compact, then $\hat{A}$ is discrete. If A is discrete, then $\hat{A}$ is compact.*

Proof: Suppose that A is compact. We then choose the exhaustion to be $K_1 = K_2 = \cdots = A$ and the metric on $\hat{A}$ to be

$$d(\chi, \eta) \;=\; \sup_{x \in A} |\chi(x) - \eta(x)|.$$

To show that $\hat{A}$ is discrete, it suffices to show that for every two characters χ and η, if $d(\chi, \eta) \le \sqrt{2}$, then $\chi = \eta$. For this let $\alpha = \chi^{-1}\eta$ and assume $d(\alpha, 1) \le \sqrt{2}$. This means that

$$\alpha(A) \;\subset\; \{\mathrm{Re}(z) \ge 0\}.$$

Since $\alpha(A)$ is a subgroup of $\mathbb{T}$, we infer that $\alpha(A) = \{1\}$, so $\alpha = 1$, i.e., $\chi = \eta$.

Now let A be discrete. Being σ-compact, A is countable. Let $(a_k)_{k \in \mathbb{N}}$ be an enumeration of A. Let χ_j be a sequence in $\hat{A}$. As in the proof of Theorem 7.1.4 we find a subsequence χ_j^j of χ_j such that all sequences $\chi_j^j(a_k)$ converge. But this just means that χ_j^j converges pointwise, and hence locally uniformly. Thus the limit is a character again, and so $\hat{A}$ is compact. $\qquad\square$

The examples of Proposition 7.1.1 suggest that the bidual $\hat{\hat{A}}$ should coincide with A. Indeed, we have the following theorem.

Theorem 7.2.2 *(Pontryagin Duality) Let A denote an LCA group. Then the map*

$$A \;\to\; \hat{\hat{A}},$$
$$a \;\mapsto\; \delta_a, \qquad \delta_a(\chi) = \chi(a),$$

is an isomorphism of LCA groups.

The proof relies on deep structure theorems on LCA groups and will not be given here. The interested reader is referred to [11].

Note that the duality between $\mathbb{Z}$ and $\mathbb{R}/\mathbb{Z}$ and the self-duality of $\mathbb{R}$ proves the theorem for these groups.

7.3   Exercises

Exercise 7.1 Let K be a compact group. Show that there is no continuous group homomorphism $\eta : K \to \mathbb{R}$ except the trivial one.

Exercise 7.2 Let A and B be LCA groups. Show that there is an isomorphism of LCA groups
$$\widehat{A \times B} \cong \hat{A} \times \hat{B}.$$

Exercise 7.3 Let $\mathbb{C}^\times$ denote the multiplicative group of the complex numbers without zero. Show that $\mathbb{C}^\times$ is an LCA group and show that
$$\widehat{\mathbb{C}^\times} \cong \mathbb{Z} \times \mathbb{R}.$$

Exercise 7.4 Let A be an LCA group. Show that there is a sequence V_n of neighborhoods of the unit element e such that
$$\overline{V_{n+1}^2} = \overline{\{uv | u, v \in V_{n+1}\}} \subset V_n$$
and
$$\bigcap_{n \in \mathbb{N}} V_n = \{e\}.$$
(Hint: Show that the continuity of the multiplication implies that for every neighborhood V of the unit there is a neighborhood U of the unit such that $U^2 \subset V$. Now choose a metric and consider balls around the unit.)

Exercise 7.5 For $j \in \mathbb{N}$ let A_j be a finite abelian group. Consider the compact LCA group $\prod_j A_j$ (Exercise 6.19). Show that
$$\widehat{\prod_j A_j} \cong \bigoplus_j \widehat{A_j},$$
where the direct sum on the right-hand side is the set of all $\chi \in \prod_j \hat{A}_j$ with $\chi_j = 1$ for all but finitely many j, endowed with the discrete metric.

Exercise 7.6 Let $\psi : A \rightarrow B$ be a continuous homomorphism of LCA groups. For $\chi \in \hat{B}$ define $\psi^*(\chi) : A \rightarrow \mathbb{C}$ by

$$\psi^*(\chi)(a) = \chi(\psi(a)).$$

Show that ψ^* is a continuous group homomorphism $\hat{B} \rightarrow \hat{A}$.

Exercise 7.7 A metric d on an abelian group A is called invariant if

$$d(a, b) = d(ac, bc)$$

for every $a, b, c \in A$. Show that for every LCA group A there is an invariant metric in the metric class.

(Hint: Use the isomorphism $A \rightarrow \hat{\hat{A}}$.)

Exercise 7.8 Let B be a closed subgroup of the LCA group A. Show that B again is an LCA group and so is the quotient A/B.

Let res : $\hat{A} \rightarrow \hat{B}$ be the restriction. Show that the kernel of res is isomorphic to $\widehat{A/B}$.

(Hint: To construct a metric on A/B take an invariant metric on A and take the infimum of its B-translates.)

Exercise 7.9 Let $1 \rightarrow A \rightarrow B \rightarrow C \rightarrow 1$ be a short exact sequence of LCA groups with continuous homomorphisms. Show that this induces an exact sequence

$$1 \rightarrow \hat{C} \rightarrow \hat{B} \rightarrow \hat{A} \rightarrow 1.$$

Exercise 7.10 Let the notation be as in Exercise 6.20. Show that

$$\widehat{\varprojlim A_j} \cong \varinjlim \widehat{A_j}.$$

Exercise 7.11 Let the notation be as in Exercise 6.23. Show that

$$\widehat{\varinjlim B_j} \cong \varprojlim \widehat{B_j}.$$

Chapter 8

Plancherel Theorem

In this chapter the general Plancherel theorem will be given. The general Plancherel theorem is a simultaneous generalization of the completeness of Fourier series and the Plancherel theorem for the real line. Therefore, it shows how abstract harmonic analysis indeed is a generalization of Fourier analysis. To be able to formulate the general Plancherel theorem for LCA groups we first need the notion of Haar integration.

8.1   Haar Integration

In this section we seek to generalize the Riemann integral on the reals to a general LCA group. Indeed, it turns out that this construction works equally well for nonabelian groups, so we may perform it on an arbitrary metrizable, σ-locally compact group G, or an LC group for short.

Example. Let $G = \mathrm{GL}_n(\mathbb{R})$ be the group of invertible $n \times n$-matrices over $\mathbb{R}$. Since $\mathrm{GL}_n(\mathbb{R}) \subset \mathrm{Mat}_n(\mathbb{R}) \cong R^{n^2}$, we have a natural locally compact σ-compact topology on G. The group laws are given by matrix multiplication and inversion and thus are continuous.

To generalize the Riemann integral we first have to give a different description of it. So, let f be a real-valued continuous function with compact support on $\mathbb{R}$ that is nonnegative, i.e., $f(x) \geq 0$ for every $x \in \mathbb{R}$. The Riemann integral of f is given by the infimum of the integrals of Riemann step functions that dominate f. This can be stated as

112 CHAPTER 8. PLANCHEREL THEOREM

follows: For $n \in \mathbb{N}$ let $\mathbf{1}_n$ be the characteristic function of the interval $\left[-\frac{1}{2n}, \frac{1}{2n}\right]$. Then there are $x_1, \ldots, x_m \in \mathbb{R}$ and $c_1, \ldots, c_m > 0$ such that

$$f(x) \leq \sum_{j=1}^{m} c_j \mathbf{1}_n(x_j + x).$$

Let $(f : \mathbf{1}_n)$ denote the following infimum

$$\inf \left\{ \sum_{j=1}^{m} c_j \;\middle|\; \begin{array}{c} c_1, \ldots, c_m > 0, \text{ and there are } x_1, \ldots, x_m \in \mathbb{R} \\ \text{such that } f(x) \leq \sum_{j=1}^{m} c_j \mathbf{1}_n(x_j + x) \end{array} \right\}.$$

Then the Riemann integral can be described as

$$\int_{-\infty}^{\infty} f(x)dx = \lim_{n \to \infty} \frac{1}{n}(f : \mathbf{1}_n).$$

On a general LC group G we replace the interval $\left[-\frac{1}{2n}, \frac{1}{2n}\right]$ by an arbitrary neighborhood of the neutral element, but it is not immediately clear what the replacement of the length factor $1/n$ might be. Consequently, we have to alter the description of the Riemann integral yet further. Let f_0 denote the characteristic function of the interval $[0, 1]$. Then

$$(f_0 : \mathbf{1}_n) = n$$

for each $n \in \mathbb{N}$. Therefore,

$$\int_{-\infty}^{\infty} f(x)dx = \lim_{n \to \infty} \frac{(f : \mathbf{1}_n)}{(f_0 : \mathbf{1}_n)}.$$

In the case of a general LC group G, we replace $\mathbf{1}_n$ by the characteristic function of a compact neighborhood U of the neutral element and fix a function $f_0 \geq 0$ that is nonzero. We let the neighborhood U shrink to $\{e\}$ and define

$$\int_G f(x)dx = \lim_{U \to \{e\}} \frac{(f : \mathbf{1}_U)}{(f_0 : \mathbf{1}_U)}.$$

The difficulty here is to show that the limit exists.

This construction gives the existence of the so-called Haar integral. To explain this notion let G denote an LC group. Recall the *support* of a function f on a metric space X is the set

$$\mathrm{supp}(f) \stackrel{\text{def}}{=} \overline{\{x \in X \mid f(x) \neq 0\}},$$

where the overline means the closure.

Let $C_c(G)$ be the complex vector space of all continuous functions from G to $\mathbb{C}$ that have compact support. For a given complex vector space V, a linear map $L : V \to \mathbb{C}$ is also called a *linear functional* on V. We say that a function $f \in C_c(G)$ is *nonnegative*, and we write $f \geq 0$, if $f(x) \geq 0$ for every $x \in G$. A linear functional I on $C_c(G)$ is called an *integral* if for $f \in C_c(G)$,

$$f \geq 0 \quad \Rightarrow \quad I(f) \geq 0.$$

Example. Let $x \in G$ and let $\delta_x(f) = f(x)$, where $f \in C_c(G)$. Then δ_x is an integral, called the *Dirac distribution at* x.

If it is clear which integral to use, we will also write

$$I(f) \; = \; \int_G f(x) dx.$$

If $f, g \in C_c(G)$ are real-valued, we write $f \geq g$ if $f - g \geq 0$. It then follows that

$$f \geq g \quad \Rightarrow \quad I(f) \; \geq \; I(g).$$

The following lemma is often used.

Lemma 8.1.1 *For every integral on G we have*

$$\left| \int_G f(x) dx \right| \; \leq \; \int_G |f(x)| dx.$$

Proof: Note first that since the integral is $\mathbb{C}$-linear and maps real-valued functions to $\mathbb{R}$, we have $\mathrm{Re}\left(\int_G f(x) dx \right) = \int_G \mathrm{Re}(f(x)) dx$. Next, to prove the lemma, we may multiply f by a complex number θ of absolute value 1 without changing either side of the inequality in question. So we may assume that $\int_G f(x) dx$ is real. So assuming that we have proven the claim for real-valued functions, we can argue as follows:

$$\left| \int_G f(x) dx \right| \; = \; \left| \mathrm{Re}\left(\int_G f(x) dx \right) \right| \; = \; \left| \int_G \mathrm{Re}(f(x)) dx \right|$$
$$\leq \; \int_G |\mathrm{Re}(f(x))| dx \; \leq \; \int_G |f(x)| dx,$$

since $|\text{Re}(f(x))| \leq |f(x)|$. So it remains to prove the claim for a real-valued function f. Let

$$f_\pm = \max(\pm f, 0).$$

Then $f_\pm \in C_c(G)$, the function $f_\pm$ is nonnegative, one has $f = f_+ - f_-$, and $|f| = f_+ + f_-$, so that

$$
\left| \int_G f(x)dx \right| = \left| \int_G f_+(x)dx - \int_G f_-(x)dx \right|
$$

$$
\leq \left| \int_G f_+(x)dx \right| + \left| \int_G f_-(x)dx \right|
$$

$$
= \int_G f_+(x)dx + \int_G f_-(x)dx = \int_G |f(x)|dx.
$$

This finishes the proof of the lemma.   □

Let $s \in G$ and $f \in C_c(G)$. For $x \in G$ define

$$L_s f(x) = f(s^{-1}x),$$

the *left translation* by s. Then the function $L_s f$ again lies in $C_c(G)$ with $L_s(L_t f) = L_{st}f$ for $s, t \in G$ and $L_1 f = f$. Thus the group G acts on $C_c(G)$ by left translation. An integral $I : C_c(G) \to \mathbb{C}$ is called *invariant* or *left invariant* if

$$I(L_s f) = I(f)$$

holds for all $f \in C_c(G)$ and all $s \in G$. Using the above notation we note that an integral $\int_G dx$ is invariant if and only if for every $f \in C_c(G)$ and every $y \in G$ we have

$$\int_G f(yx)dx = \int_G f(x)dx.$$

Consider the example $G = \mathbb{R}$ and the linear functional

$$
\begin{aligned}
I : \quad C_c(\mathbb{R}) &\to \quad \mathbb{C}, \\
f &\mapsto \quad I(f) = \int_{-\infty}^{\infty} f(x)dx.
\end{aligned}
$$

This gives an invariant integral on $G = \mathbb{R}$, called the Riemann integral.

The aim of this chapter is to generalize this linear map I to the case of an arbitrary G as follows.

Theorem 8.1.2 *There exists a non-zero invariant integral I of G. If I' is a second non-zero invariant integral, then there is a number $c > 0$ such that $I' = cI$. Any such invariant integral is called a* Haar integral.

For the uniqueness part of the theorem we say that the invariant integral is *unique up to scaling*. A proof of this theorem is given in Appendix B.

Corollary 8.1.3 *For every non-zero invariant integral I and every $g \in C_c(G)$ with $g \geq 0$ we have that $I(g) = 0$ implies $g = 0$.*

Proof: Let $g \in C_c(G)$ with $g \geq 0$ and $g \neq 0$. We have to show that $I(G) \neq 0$. For this choose $f \in C_c(G)$ with $f \geq 0$ and $I(f) \neq 0$. Since $g \neq 0$ there exist $x_1, \ldots x_n \in G$ and $c_1, \ldots c_n > 0$ such that

$$f \leq \sum_{j=1}^{n} c_j L_{x_j} g.$$

Therefore,

$$0 < I(f) \leq \sum_{j=1}^{n} c_j I(L_{x_j} g)$$

$$= \left(\sum_{j=1}^{n} c_j \right) I(g),$$

and hence $I(g) \neq 0$. $\qquad\square$

Lemma 8.1.4 *The space $C_c(G)$ is a pre-Hilbert space with the inner product*

$$\langle f, g \rangle = \int_G f(x)\overline{g(x)}dx.$$

Proof: The only thing that needs to be established is the positive definiteness. So assume that $f \in C_c(G)$ with $\langle f, f \rangle = 0$. Then the function $|f|^2 \in C_c(G)$ is positive; hence Corollary 8.1.3 implies that $|f|^2 = 0$, which implies that $f = 0$. $\qquad\square$

The Hilbert space completion of $C_c(G)$ is called $L^2(G)$. It does not depend on the choice of the Haar integral, since a different choice only varies the inner product by a scalar.

Examples of Haar integrals

- A Haar integral on $\mathbb{R}$ is given by

$$I(f) = \int_{-\infty}^{\infty} f(x)dx.$$

- A Haar integral on $\mathbb{R}/\mathbb{Z}$ is given by

$$I(f) = \int_0^1 f(x)dx.$$

- A Haar integral on the multiplicative group $\mathbb{R}_+^{\times}$ is given by

$$I(f) = \int_0^{\infty} f(x)\frac{dx}{x}.$$

- The group $GL_n(\mathbb{R})$ of all invertible real $n \times n$ matrices is an LC group. This will not be proven until Chapter 9, but we nevertheless give the Haar integral here. It is

$$I(f) = \int_{-\infty}^{\infty} \cdots \int_{-\infty}^{\infty} f \begin{pmatrix} a_{1,1} & \cdots & a_{1,n} \\ \vdots & & \vdots \\ a_{n,1} & \cdots & a_{n,n} \end{pmatrix} \frac{da_{1,1} \cdots da_{n,n}}{|\det(a)|^n}$$

(see Exercise 8.7).

8.2   Fubini's Theorem

Let G and H be metrizable, σ-compact, locally compact groups. Then the Cartesian product $G \times H$ is a group of the same type, so it has a Haar integral. We will now show that the Haar integral of $G \times H$ is given as a product of the Haar integrals of G and H.

Theorem 8.2.1 Let $I_G(g) = \int_G g(x)dx$ be a Haar integral on G. Then for every $f \in C_c(G \times H)$ the function

$$y \mapsto I_G(f(.,y)) = \int_G f(x,y)dx$$

lies in $C_c(H)$. *Let* $I_H(h) = \int_H h(y)dy$ *be a Haar integral on* H. *Then a Haar integral on* $G \times H$ *is given by*

$$I(f) = \int_H \int_G f(x,y)dx\, dy.$$

This operation can also be performed in the opposite order, yielding the same result, so

$$\int_H \int_G f(x,y)dx\, dy = \int_G \int_H f(x,y)dy\, dx.$$

Proof: We say that a sequence g_n in $C_c(G)$ converges[1] to a function $g \in C_c(G)$ if g_n converges to g uniformly on G and there is a compact set $K \subset G$ such that $\mathrm{supp}(g_n) \subset K$ for every $n \in \mathbb{N}$. It then follows that the support of g is also contained in K.

Lemma 8.2.2 *If a sequence* g_n *converges to* g *in* $C_c(G)$, *then the sequence* $I(g_n)$ *converges to* $I(g)$.

Proof: Let $K \subset G$ be compact with $\mathrm{supp}(g_n) \subset K$ for every n, and let $\chi \in C_c(G)$ be such that $\chi \equiv 1$ on K and $\chi \geq 0$. Let $c = \int_G \chi(x)dx$. Let $\varepsilon > 0$; then there is a natural number n_0 such that for $n \geq n_0$ we have $|g_n(x) - g(x)| < \varepsilon/c$ for all $x \in G$. For $n \geq n_0$ we thus have

$$\left| \int_G g_n(x)dx - \int_G g(x)dx \right| = \left| \int_G g_n(x) - g(x)dx \right|$$
$$\leq \int_G |g_n(x) - g(x)|dx$$
$$= \int_G \chi(x)|g_n(x) - g(x)|dx$$
$$< \frac{\varepsilon}{c} \int_G \chi(x)dx = \varepsilon.$$

This implies the lemma. □

Now let $f \in C_c(G \times H)$. Let y_n be a sequence in H converging to y and let $g_n(x) = f(x, y_n)$. Since f is uniformly continuous, it follows

---

[1]The reader should be aware that this notion does not fully describe the usual topology on $C_c(G)$, which is an inductive limit topology. As long as one only considers linear maps to $\mathbb{C}$, however, it suffices to consider sequences as given here.

that g_n converges to $g(x) = f(x, y)$. This implies that the function

$$y \mapsto I_G(f(., y)) = \int_G f(x, y) dx$$

is continuous.

The projection $G \times H \to H$ is a continuous map, and so the image of the support of f is compact in H. For y outside this image we have $f(x, y) = 0$ for all $x \in G$, so $I_G(f(., y)) = 0$, and hence the function $y \mapsto I_G(f(., y))$ lies in $C_c(H)$. It thus makes sense to define

$$I_1(f) = \int_H \int_G f(x, y) dx \, dy.$$

Let $s = (x_0, y_0) \in G \times H$. Then

$$I_1(L_s f) = \int_H \int_G f(x_0^{-1} x, y_0^{-1} y) dx \, dy = \int_H \int_G f(x, y_0^{-1} y) dx \, dy$$

$$= \int_H \int_G f(x, y) dx \, dy = I_1(f),$$

so that I_1 is a Haar integral. In the same way we see that

$$I_2(f) = \int_G \int_H f(x, y) dy \, dx$$

is a Haar integral, so these two integrals differ only by a scalar. To see that this scalar actually is 1, it suffices to plug in one example of a function that lies in $C_c^+(G \times H) \smallsetminus \{0\}$. Let $g \in C_c^+(G)$ and $h \in C_c^+(H)$ be nonzero and set $f(x, y) = g(x)h(y)$; then $f \in C_c^+(G \times H)$ is nonzero and

$$I_1(f) = \int_G g(x) dx \int_H h(y) dy = I_2(f).$$

The theorem is proven.   □

The general Fubini theorem says that on a measure space one may interchange the order of integration in the case of absolute convergence. We will prove only the special case of Haar integration that suffices for our needs.

Let $f : G \to [0, \infty)$ be continuous and define

$$\int_G f(x) dx = \sup_{\substack{\varphi \in C_c(G) \\ 0 \le \varphi \le f}} \int_G \varphi(x) dx \in [0, \infty].$$

Let g be another continuous function from G to $[0, \infty)$. For $\varphi, \psi \in C_c^+(G)$ with $\varphi \leq f$ and $\psi \leq g$ it follows that $\varphi + \psi \leq f + g$, which implies that $\int_G f(x)dx + \int_G g(x)dx \leq \int_G (f(x) + g(x))dx$. Since, on the other hand, every function $\eta \in C_c^+(G)$ such that $\eta \leq f + g$ can be written as a sum $\eta = \varphi + \psi$ as above, it follows that equality holds, so

$$\int_G (f(x) + g(x))dx = \int_G f(x)dx + \int_G g(x)dx.$$

Let $L_{\mathrm{bc}}^1(G)$ be the set of all $f : G \to \mathbb{C}$ that are bounded and continuous, and satisfy

$$\|f\|_1 = \int_G |f(x)|dx < \infty.$$

Likewise, let $L_{\mathrm{bc}}^2(G)$ be the set of all bounded and continuous f that satisfy

$$\|f\|_2^2 = \int_G |f(x)|^2 dx < \infty.$$

Then $L_{\mathrm{bc}}^1(G)$ is a subset of $L_{\mathrm{bc}}^2(G)$, and both are complex vector spaces. The latter is indeed a pre-Hilbert space with scalar product

$$\langle f, g \rangle = \int_G f(x)\overline{g(x)}dx.$$

This notion, however, has to be defined. We do this as follows: Let $f \in L_{\mathrm{bc}}^1(G)$; then $f = u + iv$ for real-valued functions $u, v \in L_{\mathrm{bc}}^1(A)$. Next let $u_+(x) = \max(u(x), 0)$ and $u_-(x) = \max(-u(x), 0)$. Then the functions $u_\pm$ are nonnegative and continuous, and $u_\pm \leq |f|$, so $u_\pm \in L_{\mathrm{bc}}^1(G)$. We have $u = u_+ - u_-$. Similarly, we get $v = v_+ - v_-$, and so $f = u_+ - u_- + i(v_+ - v_-)$. Now set

$$\int_G f(x)dx = \int_G u_+dx - \int_G u_-dx + i\left(\int_G v_+dx - \int_G v_-dx\right).$$

Then $\left|\int_G f(x)dx\right| \leq \int_G |f(x)|dx$.

Now let H be another LC group; then the product $G \times H$ is also an LC group. Fix a Haar integral on H.

Lemma 8.2.3 (*Fubini's theorem, weak version*) *Let $f \in L_{\mathrm{bc}}^1(G \times H)$ and assume that the function $y \mapsto \int_G f(x, y)dx$ lies in $L_{\mathrm{bc}}^1(H)$, and the same holds with G and H interchanged. Then*

$$\int_G \int_H f(x, y)dy\,dx = \int_H \int_G f(x, y)dx\,dy.$$

Proof: This follows directly from the definitions and Theorem 8.2.1.

$\square$

8.3   Convolution

Back to abelian groups; let A be an LCA group. Fix a Haar integral $\int_A dx$. Let $\hat{A}$ be the dual group, i.e., the group of all characters $\chi : A \to \mathbb{T}$. For $f \in L^1_{bc}(A)$ let $\hat{f} : \hat{A} \to \mathbb{C}$ be its *Fourier transform* defined by

$$\hat{f}(\chi) = \int_A f(x)\overline{\chi(x)}dx.$$

This definition of the Fourier transform fits well with the previous one for the group $\mathbb{R}$ as in Section 3.3. To see this, let $x \in \mathbb{R}$ and let φ_x be the character attached to x, i.e., $\varphi_x(y) = e^{2\pi i x y}$. Then for $f \in L^1_{bc}(\mathbb{R})$ we have

$$\hat{f}(\varphi_x) = \int_{\mathbb{R}} f(y)\overline{\varphi_x(y)}dy = \int_{-\infty}^{\infty} f(y)\, e^{-2\pi i x y}dy = \hat{f}(x),$$

where the first $\hat{f}$ is the new definition of the Fourier transform and the second is the old one. This justifies the use of the same symbol here.

For the group $\mathbb{R}/\mathbb{Z}$ the dual is $\mathbb{Z}$, so the Fourier transform $\hat{f}$ is a function on $\mathbb{Z}$. For $k \in \mathbb{Z}$ we compute

$$\hat{f}(k) = \int_{\mathbb{R}/\mathbb{Z}} f(y)\, e^{-2\pi i k y}dy = c_k(f).$$

So, in the case of $\mathbb{R}/\mathbb{Z}$ the abstract Fourier transform is given simply by taking the kth Fourier coefficient. In this way we see how the theory of the abstract Fourier transform generalizes both the theory of Fourier series and the theory of the Fourier transform on the reals.

Theorem 8.3.1 *Let $f, g \in L^1_{bc}(A)$. Then the integral*

$$f * g(x) = \int_A f(xy^{-1})g(y)dy$$

exists for every $x \in A$ and defines a function $f * g \in L^1_{bc}(A)$. For the Fourier transform we have

$$\widehat{f * g}(\chi) = \hat{f}(\chi)\hat{g}(\chi)$$

for every $\chi \in \hat{A}$.

Note that for the group $\mathbb{R}$ this definition of the convolution coincides with the one in Section 3.2 and that the second assertion of the present theorem generalizes Theorem 3.3.1, part c.

Proof: Assume that $|f(x)| \leq C$ for every $x \in A$. Then

$$\int_A |f(xy^{-1})g(y)|dy \leq C \int_A |g(y)|dy = C \|g\|_1 \,,$$

so the integral exists and the function $f * g$ is bounded. Next we shall prove that it is continuous. Let $x_0 \in A$. Assume $|f(x)|, |g(x)| \leq C$ for all $x \in A$ and assume $g \neq 0$. For a given $\varepsilon > 0$ there is a function $\varphi \in C_c^+(A)$ such that $\varphi \leq |g|$ and

$$\int_A |g(y)| - \varphi(y)dy < \frac{\varepsilon}{4C}.$$

On a compact set the function f is uniformly continuous, so there is a neighborhood V of the unit element such that $x \in Vx_0$, $y \in \text{supp}\varphi$ implies $|f(xy^{-1}) - f(x_0y^{-1})| < \varepsilon/2 \|g\|_1$. It follows that for $x \in Vx_0$,

$$\int_A |f(xy^{-1}) - f(x_0y^{-1})|\varphi(y)dy \leq \frac{\varepsilon}{2 \|g\|_1} \int_A \varphi(y)\,dy \leq \frac{\varepsilon}{2},$$

and on the other hand,

$$\int_A |f(xy^{-1}) - f(x_0y^{-1})|(|g(y)| - \varphi(y))dy$$

is less than or equal to

$$2C \int_A |g(y)| - \varphi(y)dy < \frac{\varepsilon}{2},$$

so that for $x \in x_0V$,

$$
\begin{aligned}
|f * g(x) - f * g(x_0)| &= \left| \int_A (f(xy^{-1}) - f(x_0y^{-1}))g(y)dy \right| \\
&\leq \int_A |f(xy^{-1}) - f(x_0y^{-1})||g(y)|dy \\
&= \int_A |f(xy^{-1}) - f(x_0y^{-1})| \\
&\qquad \times ((|g(y)| - \varphi(y)) + \varphi(y))dy \\
&\leq \frac{\varepsilon}{2} + \frac{\varepsilon}{2} = \varepsilon.
\end{aligned}
$$

Thus the function $f * g$ is continuous at x_0. To see that $\|f * g\|_1 < \infty$ we compute

$$
\begin{aligned}
\|f * g\|_1 &= \int_A |f * g(x)| dx = \int_A \left| \int_A f(xy^{-1}) g(y) dy \right| dx \\
&\leq \int_A \int_A |f(xy^{-1}) g(y)| dy dx \\
&= \int_A \int_A |f(xy^{-1}) g(y)| dx dy \\
&= \int_A |f(x)| dx \int_A |g(y)| dy = \|f\|_1 \|g\|_1 ,
\end{aligned}
$$

where we have applied Fubini's theorem and used the invariance of the Haar integral. Finally, for the Fourier transform we compute

$$
\begin{aligned}
\widehat{f * g}(\chi) &= \int_A f * g(x) \overline{\chi(x)} dx \\
&= \int_A \int_A f(xy^{-1}) g(y) \overline{\chi(x)} dy \, dx \\
&= \int_A \int_A f(y^{-1}x) g(y) \overline{\chi(x)} dx \, dy \\
&= \int_A \int_A f(x) g(y) \overline{\chi(yx)} dx \, dy \\
&= \int_A f(x) \overline{\chi(x)} dx \int_A g(y) \overline{\chi(y)} dy \\
&= \hat{f}(\chi) \hat{g}(\chi).
\end{aligned}
$$

The theorem is proven.   □

8.4   Plancherel's Theorem

The following lemma will be needed in the sequel.

Lemma 8.4.1 *Let A be a compact abelian group. Fix a Haar integral such that*

$$ \int_A 1 dx = 1. $$

Then, for every two characters $\chi, \eta \in \hat{A}$ we have

$$ \int_A \chi(x) \overline{\eta(x)} dx = \begin{cases} 1 & \text{if } \chi = \eta, \\ 0 & \text{otherwise.} \end{cases} $$

Proof: If $\chi = \eta$, then $\chi(x)\overline{\eta(x)} = 1$, so the claim follows in this case. Now suppose $\chi \neq \eta$. Then $\alpha = \chi\overline{\eta} = \chi\eta^{-1} \neq 1$, so there is $a \in A$ with $\alpha(a) \neq 1$. Then

$$\alpha(a) \int_A \alpha(x)dx = \int_A \alpha(ax)dx = \int_A \alpha(x)dx$$

by the invariance of the Haar integral. Therefore,

$$(\alpha(a) - 1) \int_A \alpha(x)dx = 0,$$

which implies

$$\int_A \alpha(x)dx = 0.$$

$\square$

Theorem 8.4.2 *Let A be an LCA group. There is a unique Haar measure on $\hat{A}$ such that for every $f \in L^1_{bc}(A)$,*

$$\|f\|_2 = \left\|\hat{f}\right\|_2;$$

i.e., for $f \in L^1_{bc}(A)$ the Fourier transform $\hat{f}$ lies in $L^2_{bc}(\hat{A})$ and the Fourier transform extends to a Hilbert space isomorphism of the completions $L^2(A) \to L^2(\hat{A})$.

This is the point at which it becomes transparent how abstract harmonic analysis indeed generalizes the theory of Fourier series and the Fourier transform on the reals. If we specialize the above theorem to the case of the group $\mathbb{R}/\mathbb{Z}$, we get for $f \in L^1_{bc}(\mathbb{R}/\mathbb{Z})$,

$$\int_0^1 |f(x)|^2 dx = \|f\|_2^2 = \left\|\hat{f}\right\|_2^2 = \sum_{k \in \mathbb{Z}} |\hat{f}(k)|^2 = \sum_{k \in \mathbb{Z}} |c_k(f)|^2.$$

Modulo the easy Lemma 1.3.1 this result implies the completeness of the Fourier series (Theorem 1.4.4). The present theorem also is a generalization of Plancherel's theorem for the real Fourier transform (Theorem 3.5.2) in an even more obvious fashion.

Proof of the Theorem: The proof of this theorem in full generality is beyond our scope. The interested reader is referred to [11]. We will here prove the result only for the special case of a discrete group A. As Haar integral we choose

$$\int_A f(x)dx = \sum_{a \in A} f(a).$$

For $\chi \in \hat{A}$ we then have $\hat{f}(\chi) = \sum_{a \in A} f(a)\overline{\chi(a)}$. On $\hat{A}$ we choose the Haar integral normalized by the condition $\int_{\hat{A}} 1 da = 1$. Then Lemma 8.4.1 applies to $\hat{A}$.

Lemma 8.4.3 *For every $g \in L^1_{bc}(A)$ the Fourier transform $\hat{g}$ is in $C(\hat{A}) = L^1_{bc}(\hat{A})$, and we have, for every $a \in A$,*

$$\hat{\hat{g}}(\delta_a) = g(a^{-1}).$$

Proof: We compute

$$\hat{\hat{g}}(\delta_a) = \int_{\hat{A}} \hat{g}(\chi)\overline{\delta_a(\chi)}d\chi = \int_{\hat{A}} \sum_{b \in A} g(b)\overline{\delta_b(\chi)}\overline{\delta_a(\chi)}d\chi$$

$$= \int_{\hat{A}} \sum_{b \in A} g(b^{-1})\delta_b(\chi)\overline{\delta_a(\chi)}d\chi$$

$$= \sum_{b \in A} g(b^{-1}) \int_{\hat{A}} \delta_b(\chi)\overline{\delta_a(\chi)}d\chi = g(a^{-1})$$

according to Lemma 8.4.1, since the δ_a are precisely the characters of $\hat{A}$ by duality. The lemma follows. $\square$

To prove the theorem in the discrete case, let $f \in L^1_{bc}(A)$ and set $\tilde{f}(x) = \overline{f(x^{-1})}$. Set $g = \tilde{f} * f$. Then

$$g(x) = \int_A \overline{f(yx^{-1})}f(y)dy,$$

so that $g(e) = \|f\|_2^2$, where e is the unit element in A. By Theorem 8.3.1 we have $\hat{g}(\chi) = \hat{\tilde{f}}(\chi)\hat{f}(\chi) = \overline{\hat{f}(\chi)}\hat{f}(\chi) = |\hat{f}(\chi)|^2$. We get

$$\|f\|_2^2 = g(e) = \hat{\hat{g}}(\delta_e) = \int_{\hat{A}} \hat{g}(\chi)\overline{\chi(e)}d\chi$$

$$= \int_{\hat{A}} |\hat{f}(\chi)|^2 d\chi = \|\hat{f}\|_2^2.$$

This proves the claim for A discrete. $\square$

8.5 Exercises

Exercise 8.1 Let G be a discrete group. Show that

$$I(f) = \sum_{x \in G} f(x)$$

is well-defined for $f \in C_c(G)$ and defines a Haar integral for G.

Exercise 8.2 Show that for every open set $V \subset G$ there is a function $\varphi \in C_c(G)$ that is nonzero and satisfies $\operatorname{supp}(\varphi) \subset V$. Show that for every compact subset $K \subset G$ there is a function $\chi \in C_c(G)$ such that $\chi \equiv 1$ on K.

(Hint: Fix a metric d and show that the function $x \mapsto d(x_0, x)$ is continuous for given $x_0 \in G$.)

Exercise 8.3 Show that every $f \in C_c(G)$ is uniformly continuous; i.e., for every $\varepsilon > 0$ there is a neighborhood V of the unit element such that for each $x, y \in G$ we have

$$x^{-1}y \in V \quad \Rightarrow \quad |f(x) - f(y)| < \varepsilon.$$

Exercise 8.4 Let B be the subgroup of $\operatorname{GL}_2(\mathbb{R})$ defined as

$$B = \left\{ \begin{pmatrix} 1 & b \\ & c \end{pmatrix} \middle| b, c \in \mathbb{R},\ c \neq 0 \right\}.$$

Show that

$$I(f) = \int_{\mathbb{R}^\times} \int_{\mathbb{R}} f\left(\begin{pmatrix} 1 & b \\ & c \end{pmatrix} \right) db \frac{dc}{c}$$

is a Haar integral on B. Show that I is not right invariant; i.e., there are $z \in B$ and $f \in C_c(B)$ such that $I(R_z f) \neq I(f)$, where $R_z f(x) = f(xz)$.

Exercise 8.5 Let G be an LC group with Haar integral. Show that for $x \in G$ the map

$$f \mapsto \int_G f(xy) dy$$

is also a Haar integral. Conclude from the uniqueness of the Haar integral that there is a function $\Delta : G \to \mathbb{R}_+^\times$ such that

$$\int_G f(xy) dx = \Delta(y) \int_G f(x) dx$$

holds for every $f \in C_c(G)$. Show that Δ is a continuous group homomorphism. The function Δ is called the *modular function* of G. Show that Δ is trivial if and only if the Haar integral of G is also right invariant. In this case G is called *unimodular*.

Exercise 8.6 Show that a compact group K is unimodular and infer that

$$\int_K f(k)dk = \int_K f(k^{-1})dk$$

for every $f \in C(K)$.

(Hint: Show that the image of the modular function is trivial.)

Exercise 8.7 Show that a Haar integral for the group $\mathrm{GL}_2(\mathbb{R})$ is given by

$$I(f) = \int_{-\infty}^{\infty} \int_{-\infty}^{\infty} \int_{-\infty}^{\infty} \int_{-\infty}^{\infty} f\begin{pmatrix} x & y \\ z & w \end{pmatrix} \frac{dx\,dy\,dz\,dw}{|xw - yz|^2}.$$

Exercise 8.8 Prove that for a general LCA group A the convolution $*$ satisfies the following identities: $f * g = g * f$, $f * (g * h) = (f * g) * h$, and $f * (g + h) = f * g + f * h$ for all $f, g, h \in L^1_{bc}(A)$.

Exercise 8.9 Show that the Hilbert space $L^2(A)$ is the completion of the pre-Hilbert space $L^2_{bc}(A)$. Recall here that $L^2(A)$ is defined as the completion of $C_c(A)$.

Part III

Noncommutative Groups

Chapter 9

Matrix Groups

Matrix groups like $GL_n(\mathbb{C})$ and $U(n)$ are the most important non-commutative topological groups, since they occur naturally as transformation groups in various contexts.

9.1 $GL_n(\mathbb{C})$ and $U(n)$

Let n be a natural number. On the vector space of complex $n \times n$ matrices $\text{Mat}_n(\mathbb{C})$ we define a norm:

$$\|A\|_1 = \sum_{i,j=1}^{n} |a_{i,j}|,$$

where $A = (a_{i,j})$. This norm gives rise to the metric $d_1(A, B) = \|A - B\|_1$. On the other hand, on the vector space $\text{Mat}_n(\mathbb{C}) \cong \mathbb{C}^{n^2}$ we have a natural inner product that gives rise to a second norm, called the Euclidean norm,

$$\|A\|_2 = \sqrt{\sum_{i,j=1}^{n} |a_{i,j}|^2},$$

and we get a corresponding metric $d_2(A, B) = \|A - B\|_2$.

Lemma 9.1.1 *A sequence of matrices $A^{(k)} = (a_{i,j}^{(k)})$ converges in d_1 if and only if for each pair of indices (i, j), the sequence of entries $a_{i,j}^{(k)}$ converges in $\mathbb{C}$. The same holds for d_2, so the metrics d_1 and d_2 are equivalent.*

Proof: Suppose the sequence $A^{(k)} = (a_{i,j}^{(k)})$ converges in d_1 to $A = (a_{i,j}) \in \text{Mat}_n(\mathbb{C})$. Then for every $\varepsilon > 0$ there is $k_0 \in \mathbb{N}$ such that for all $k \geq k_0$

$$\left\| A^{(k)} - A \right\|_1 < \varepsilon.$$

Let i_0, j_0 be in $\{1, 2, \ldots, n\}$; then it follows that for $k \geq k_0$,

$$|a_{i_0,j_0}^{(k)} - a_{i_0,j_0}| \leq \sum_{i,j} |a_{i,j}^{(k)} - a_{i,j}| = \left\| A^{(k)} - A \right\|_1 < \varepsilon.$$

Therefore, each entry converges. Conversely, assume that $a_{i,j}^{(k)} \to a_{i,j}$ for each pair of indices (i, j). Then for a given $\varepsilon > 0$ there is $k_0(i, j)$ such that for $k \geq k_0(i, j)$,

$$\left| a_{i,j}^{(k)} - a_{i,j} \right| < \frac{\varepsilon}{n^2}.$$

Let $k_0 \in \mathbb{N}$ be the maximum of all $k_0(i, j)$ as (i, j) varies. Then for $k \geq k_0$,

$$\left\| A^{(k)} - A \right\|_1 = \sum_{i,j} |a_{i,j}^{(k)} - a_{i,j}| < \sum_{i,j} \frac{\varepsilon}{n^2} = \varepsilon,$$

so $A^{(k)}$ converges to A in d_1. The case of d_2 is similar. $\square$

Proposition 9.1.2 *With the topology or metric class given above, the group of complex invertible matrices, $\text{GL}_n(\mathbb{C})$, is an LC group; i.e., it is a metrizable, σ-compact, locally compact group.*

Proof: First note that multiplication and inversion are given as rational functions in the entries. Since polynomials are continuous, it follows that $\text{GL}_n(\mathbb{C})$ is a topological group. Being an open subset of the locally compact space $\text{Mat}_n(\mathbb{C})$ it is locally compact. Finally, to see that it is σ-compact, for $n \in \mathbb{N}$ let

$$K_n = \{a \in \text{GL}_n(\mathbb{C}) : \|a\|_1 \leq n, \ \left\|a^{-1}\right\|_1 \leq n\}.$$

Then every sequence in K_n must have a convergent subsequence in the finite-dimensional vector space $\text{Mat}_n(\mathbb{C})$. This convergent subsequence then must have a subsequence for which the inverses also converge, so the limit is invertible. $\square$

For $A \in \mathrm{Mat}_n(\mathbb{C})$ let A^* be its *adjoint matrix*; i.e., if $A = (a_{i,j})$, then $A^* = (\overline{a_{j,i}})$, so $A^* = \bar{A}^t$, where the bar means the complex conjugate and $(.)^t$ gives the transpose of a matrix. Let

$$U(n) = \{g \in \mathrm{Mat}_n(\mathbb{C}) \mid g^*g = 1\},$$

where 1 means the unit matrix.

Lemma 9.1.3 $U(n)$ *is a compact subgroup of* $\mathrm{GL}_n(\mathbb{C})$.

Proof: For $g \in \mathrm{Mat}_n(\mathbb{C})$ the equation $g^*g = 1$ implies that g is invertible and $g^* = g^{-1}$, so in particular, $U(n)$ is a subset of $\mathrm{GL}_n(\mathbb{C})$. Let $a, b \in U(n)$. To see that $U(n)$ is a subgroup we have to show that $ab \in U(n)$ and $a^{-1} \in U(n)$. For the first part consider $(ab)^*ab = b^*a^*ab = b^*b = 1$, so $ab \in U(n)$. For the second part recall that $a^* = a^{-1}$ implies $1 = aa^* = (a^*)^*a^*$, so $a^* = a^{-1}$ also lies in $U(n)$.

To see that $U(n)$ is compact, it suffices to show that the group $U(n)$ is closed in $\mathrm{Mat}_n(\mathbb{C})$ and bounded in the Euclidean norm (see Exercise 6.11). So let g_j be a sequence in $U(n)$ converging to g in $\mathrm{Mat}_n(\mathbb{C})$. Then

$$1 = \lim_{j\to\infty} g_j^*g_j = \left(\lim_{j\to\infty} g_j\right)^* \lim_{j\to\infty} g_j = g^*g.$$

This implies that $U(n)$ is closed. Moreover it is bounded, since for every $a \in \mathrm{Mat}_n(\mathbb{C})$ we have

$$\mathrm{tr}\,(a^*a) = \sum_{k=1}^{n}(a^*a)_{k,k} = \sum_{k=1}^{n}\sum_{j=1}^{n} a_{kj}^* a_{j,k}$$

$$= \sum_{k=1}^{n}\sum_{j=1}^{n} \overline{a_{j,k}} a_{j,k} = \sum_{k=1}^{n}\sum_{j=1}^{n} |a_{j,k}|^2 = \|a\|_2^2.$$

Therefore, $\|g\|_2 = \sqrt{\mathrm{tr}\,1} = \sqrt{n}$ for $g \in U(n)$, so $U(n)$ is bounded. $\qquad\square$

9.2 Representations

The role of characters for LCA groups will in the case of noncommutative groups be played by representations, a notion introduced in this section. Let G be a (metrizable) topological group. Let $(V, \langle ., . \rangle)$

be a Hilbert space. Let $\mathrm{GL}(V)$ be the set of all invertible linear maps $T : V \to V$. A *representation* of G on V is a group homomorphism $\eta : G \to \mathrm{GL}(V)$ such that the map

$$
\begin{aligned}
G \times V &\to V, \\
(x, v) &\mapsto \eta(x)v,
\end{aligned}
$$

is continuous. The representation η is called *unitary* if for every $x \in G$ the operator $\eta(x)$ is unitary on V, i.e., if

$$
\langle \eta(x)v, \eta(x)w \rangle = \langle v, w \rangle \quad \text{for all } v, w \in V, \ x \in G.
$$

A closed subspace $W \subset V$ is called *invariant* for η if $\eta(x)W \subset W$ for every $x \in G$. The representation η is called *irreducible* if there is no proper closed invariant subspace, i.e., the only closed invariant subspaces are 0 and V itself.

Example. The identity map $\rho : \mathrm{U}(n) \to \mathrm{GL}(\mathbb{C}^n) = \mathrm{GL}_n(\mathbb{C})$ is a unitary representation.

Lemma 9.2.1 ρ *is irreducible.*

Proof: By definition $\mathrm{U}(n)$ consists of all linear operators on $\mathbb{C}^n$ that are unitary with respect to the standard inner product $\langle v, w \rangle = v^t \bar{w}$.

Let $V \subset \mathbb{C}^n$ be a subspace which is neither zero nor the whole space. Let $W = V^\perp$ be its orthogonal space, i.e.,

$$
W = \{ w \in \mathbb{C}^n \mid \langle w, v \rangle = 0 \text{ for every } v \in V \}.
$$

Then $\mathbb{C}^n = V \oplus W$. Let $e_1, \dots, e_l$ be an orthonormal basis of V and $e_{l+1}, \dots, e_n$ be an orthonormal basis of W; then the operator T given by

$$
\begin{aligned}
T(e_1) &= e_{l+1}, \quad T(e_{l+1}) = e_1, \\
T(e_j) &= e_j \quad \text{for } j \neq 1, l+1,
\end{aligned}
$$

is unitary by Exercise 2.2. Thus $T \in \mathrm{U}(n)$, but T does not leave V stable. So there is no nontrivial invariant subspace. $\square$

9.3 The Exponential

A series in $\text{Mat}_n(\mathbb{C})$ of the form $\sum_{\nu=0}^{\infty} A_\nu$ converges by definition if the sequence of partial sums $s_k = \sum_{\nu=1}^{k} A_\nu$ converges.

Proposition 9.3.1 *For every $A \in \text{Mat}_n(\mathbb{C})$ the series*

$$\exp(A) = \sum_{\nu=0}^{\infty} \frac{A^\nu}{\nu!}$$

converges and defines an element in $\text{GL}_n(\mathbb{C})$. If $A, B \in \text{Mat}_n(\mathbb{C})$ satisfy $AB = BA$, then $\exp(A+B) = \exp(A)\exp(B)$. In particular, it follows that

$$\exp(-A) = \exp(A)^{-1}.$$

Proof: Recall the 1-norm on $\text{Mat}_n(\mathbb{C})$:

$$\|A\|_1 = \sum_{i,j=1}^{n} |a_{i,j}|$$

if $A = (a_{i,j})$.

Lemma 9.3.2 *For $A, B \in \text{Mat}_n(\mathbb{C})$ we have*

$$\|AB\|_1 \leq \|A\|_1 \|B\|_1.$$

In particular, for $j \in \mathbb{N}$, $\left\|A^j\right\|_1 \leq \|A\|_1^j$.

Proof: Let $A = (a_{i,j})$ and $B = (b_{i,j})$; then

$$\|AB\|_1 = \sum_{i,j=1}^{n} \left| \sum_{k=1}^{n} a_{i,k}b_{k,j} \right| \leq \sum_{i,j,k=1}^{n} |a_{i,k}b_{k,j}|$$

$$\leq \sum_{i,j,k,l=1}^{n} |a_{i,k}||b_{l,j}| = \|A\|_1 \|B\|_1.$$

$\square$

Lemma 9.3.3 *Let $(A_\nu)_{\nu \geq 0}$ be a sequence of matrices in $\text{Mat}_n(\mathbb{C})$. Suppose that $\sum_{\nu=0}^{\infty} \|A_\nu\|_1 < \infty$. Then the series $\sum_{\nu=0}^{\infty} A_\nu$ converges in $\text{Mat}_n(\mathbb{C})$.*

CHAPTER 9. MATRIX GROUPS

Proof: Let $B_k = \sum_{\nu=0}^{k} A_\nu$. We have to show that the sequence (B_k) converges. It suffices to show that it is a Cauchy sequence with respect to $\|.\|_1$. The sequence $b_k = \sum_{\nu=0}^{k} \|A_\nu\|_1$ converges in $\mathbb{R}$ and hence is Cauchy. So, for given $\varepsilon > 0$ there is k_0 such that for $m \geq k \geq k_0$ we have

$$\varepsilon > |b_m - b_k| = \sum_{\nu=k+1}^{m} \|A_\nu\|_1 \geq \left\| \sum_{\nu=k+1}^{m} A_\nu \right\|_1 = \|B_m - B_k\|_1 .$$

Thus it follows that (B_k) is a Cauchy sequence in $\mathrm{Mat}_n(\mathbb{C})$ and hence converges (see Exercise 9.4). $\qquad\square$

To prove the proposition it remains to show that $\sum_{\nu=0}^{\infty} \frac{\|A^\nu\|_1}{\nu!} < \infty$. We have

$$\sum_{\nu=0}^{\infty} \frac{\|A^\nu\|_1}{\nu!} \leq \sum_{\nu=0}^{\infty} \frac{(\|A\|_1)^\nu}{\nu!} < \infty,$$

since the exponential series converges in $\mathbb{R}$. The first part of the proposition follows from this. For the remainder let $A, B \in \mathrm{Mat}_n(\mathbb{C})$ with $AB = BA$. Then

$$
\begin{aligned}
\exp(A+B) &= \sum_{\nu=0}^{\infty} \frac{(A+B)^\nu}{\nu!} = \sum_{\nu=0}^{\infty} \frac{1}{\nu!} \sum_{k=0}^{\nu} \binom{\nu}{k} A^k B^{\nu-k} \\
&= \sum_{\nu=0}^{\infty} \sum_{k=0}^{\nu} \frac{1}{k!(\nu-k)!} A^k B^{\nu-k} = \exp(A)\exp(B).
\end{aligned}
$$

This implies the lemma. $\qquad\square$

Proposition 9.3.4 *For every $A \in \mathrm{Mat}_n(\mathbb{C})$ we have*

$$\det(\exp(A)) = \exp(\mathrm{tr}\,(A)).$$

Proof: Let $S \in \mathrm{GL}_n(\mathbb{C})$. Then

$$\det(\exp(SAS^{-1})) = \det(S\exp(A)S^{-1}) = \det(\exp(A))$$

and

$$\exp(\mathrm{tr}\,(SAS^{-1})) = \exp(\mathrm{tr}\,(A)),$$

so both sides in the statement of the proposition are invariant under conjugation. By the Jordan normal-form theorem every square matrix is conjugate to an upper triangular matrix, so it suffices to prove the proposition for an upper triangular matrix A. Suppose that

$$A = \begin{pmatrix} a_1 & & * \\ & \ddots & \\ & & a_n \end{pmatrix}.$$

Then for $\nu \geq 0$,

$$A^\nu = \begin{pmatrix} a_1^\nu & & * \\ & \ddots & \\ & & a_n^\nu \end{pmatrix},$$

so that

$$\exp(A) = \begin{pmatrix} e^{a_1} & & * \\ & \ddots & \\ & & e^{a_n} \end{pmatrix},$$

which gives

$$\det(\exp(A)) = e^{a_1} \cdots e^{a_n} = e^{a_1 + \cdots + a_n} = \exp(\operatorname{tr}(A)).$$

$\square$

Let $G \subset \operatorname{GL}_n(\mathbb{C})$ be a closed subgroup. The *Lie algebra* of G is by definition

$$\operatorname{Lie}(G) = \{X \in \operatorname{Mat}_n(\mathbb{C}) \mid \exp(tX) \in G \text{ for every } t \in \mathbb{R}\}.$$

Examples.

- The special linear group $\operatorname{SL}_n(\mathbb{C})$ is the group consisting of all matrices A in $\operatorname{Mat}_n(\mathbb{C})$ satisfying $\det(A) = 1$. Its Lie algebra is
$$sl_n(\mathbb{C}) = \{X \in \operatorname{Mat}_n(\mathbb{C}) \mid \operatorname{tr}(X) = 0\}.$$

- The Lie algebra of the unitary group $U(n)$ is
$$u(n) = \{X \in \operatorname{Mat}_n(\mathbb{C}) \mid X^* = -X\},$$

where $X^* = \overline{X}^t$ denotes the adjoint matrix.

In order to proceed we will need to establish some facts that are not hard to prove but the proofs require some concepts from differential geometry that are beyond the scope of this book. The following proposition will therefore not be proved here.

Proposition 9.3.5 *Let G be a closed subgroup of* $\mathrm{GL}_n(\mathbb{C})$. *Then* $\mathrm{Lie}(G)$ *is a real sub-vector space of* $\mathrm{Mat}_n(\mathbb{C})$. *If X and Y are elements of* $\mathrm{Lie}(G)$, *then so is*

$$[X, Y] \overset{\mathrm{def}}{=} XY - YX.$$

this is called the Lie bracket of X and Y. Let $\pi : G \to \mathrm{GL}(V)$ *be a finite-dimensional representation. Then for every* $X \in \mathrm{Lie}(G)$ *the map*

$$t \mapsto \pi(\exp(tX)), \quad t \in \mathbb{R},$$

is infinitely differentiable. Set

$$\pi(X) = \left. \frac{d}{dt} \right|_{t=0} \pi(\exp(tX)) \in \mathrm{End}(V).$$

Then the map $X \mapsto \pi(X)$ *is linear on* $\mathrm{Lie}(G)$ *and satisfies*

$$\pi([X, Y]) = [\pi(X), \pi(Y)],$$

where on the right-hand side we take the commutator bracket in $\mathrm{End}(V)$. *We say that* π *is a Lie algebra representation of* $\mathrm{Lie}(G)$.

Proof: The proposition follows from the material in [9], Chapter II. See also Exercise 9.14.   □

A closed subgroup G of $\mathrm{GL}_n(\mathbb{C})$ is called *path connected* if every two points $x, y \in G$ can be joined by a continuous curve, i.e., if there is a continuous map $\gamma : [0, 1] \to G$ with $\gamma(0) = x$ and $\gamma(1) = y$. For example, the multiplicative group $\mathbb{R}^\times = \mathrm{GL}_1(\mathbb{R})$ is not path connected.

Lemma 9.3.6 *If the group G is path connected and if* (π, V) *is irreducible as a representation of the group G, then it is irreducible as a representation of the Lie algebra* $\mathrm{Lie}(G)$, *i.e., there is no proper subrepresentation. Further, if G is path connected and if* π *and* π' *are isomorphic as representations of the Lie algebra, then they are isomorphic as G-representations.*

Proof: It is a consequence of Taylor's formula that for each X in the Lie algebra of G we have

$$\pi(\exp(X)) = \pi \left(\sum_{\nu=0}^{\infty} \frac{X^\nu}{\nu!} \right) = \sum_{\nu=0}^{\infty} \frac{\pi(X)^\nu}{\nu!} = \exp(\pi(X)).$$

This implies that a nonzero subspace W of V that is invariant under the Lie algebra is also invariant under the image of exp. The differential equation

$$\frac{d}{dt} \exp(tX) = X \exp(tX),$$

which follows from the series representation, implies that the differential of exp at zero is invertible, and hence the image of $\exp : \text{Lie}(G) \to G$ contains an open neighborhood of the unit. The subgroup generated by this neighborhood is an open subgroup that stabilizes W. If G is path connected, there is only one open subgroup, namely G itself (see Exercise 9.10), and hence W is stabilized by G. If π is irreducible as a G-representation, it follows that $W = V$, and so π is irreducible as a representation of the Lie algebra.

For the last point, assume that we are given two G-representations π and π' and a $\text{Lie}(G)$-isomorphism $T : V_\pi \to V_{\pi'}$; i.e., we have

$$T\pi(X) = \pi'(X)T$$

for every $X \in \text{Lie}(G)$. Since T is a linear map between finite-dimensional spaces, it is continuous, so for $X \in \text{Lie}(G)$,

$$
\begin{aligned}
T\pi(\exp(X)) &= T\exp(\pi(X)) = T\left(\sum_{\nu=0}^{\infty} \frac{\pi(X)^\nu}{\nu!}\right) \\
&= \sum_{\nu=0}^{\infty} \frac{T\pi(X)^\nu}{\nu!} = \sum_{\nu=0}^{\infty} \frac{\pi'(X)^\nu T}{\nu!} \\
&= \left(\sum_{\nu=0}^{\infty} \frac{\pi'(X)^\nu}{\nu!}\right)T = \exp(\pi'(X))T \\
&= \pi'(\exp(X))T,
\end{aligned}
$$

so T commutes with the action of the subgroup generated by the image of exp, and again, if G is path connected, this is the entire group G. Thus T is a G-isomorphism. $\qquad\square$

A representation $\pi : \text{Lie}(G) \to \text{End}(V)$ of the Lie algebra of G is called a *-representation* if for every $X \in \text{Lie}(G)$ we have

$$\pi(X)^* = \pi(-X),$$

where the $*$ refers to the adjoint in $\text{End}(V)$.

Lemma 9.3.7 *If the representation* $\pi : G \to \mathrm{GL}(V)$ *is unitary, then the derived representation of the Lie algebra is a* $*$-representation. *If* G *is path connected, then the converse is also true; i.e., if* π *is a* $*$-representation *the Lie algebra, then it is a unitary representation of the group.*

Proof: Suppose π is unitary. Then for every $x \in G$ we have $\pi(x)^* = \pi(x)^{-1} = \pi(x^{-1})$. Let $X \in \mathrm{Lie}(G)$. Then

$$
\pi(X)^* = \left(\frac{d}{dt} \pi(\exp(tX))|_{t=0} \right)^* = \frac{d}{dt} \pi(\exp(tX))^*|_{t=0}
$$

$$
= \frac{d}{dt} \pi(\exp(-tX))|_{t=0} = \pi(-X).
$$

For the converse use the equation $\pi(\exp(X)) = \exp(\pi(X))$ to see that if π is a $*$-representation, then $\pi(x)^* = \pi(x^{-1})$ for every x in the image of exp. This equation then also holds for the group generated by this image, and if G is path connected, this group equals G. $\qquad\square$

Matrix groups as featured in this chapter are special cases of Lie groups. For a nice account of Lie groups for beginners see [24].

9.4 Exercises

Exercise 9.1 Let V be a finite-dimensional Hilbert space. A linear operator $A : V \to V$ with $AA^* = A^*A$ is called *normal*. Show that every normal operator A on V is diagonalizable, i.e., there exists a basis of V consisting of A-eigenvectors.

(Hint: Use induction on the dimension. Pick an eigenspace of A and show that its orthocomplement is also invariant under A.)

Exercise 9.2 Show that the group $\mathbb{R}^\times$ is not path connected.

Exercise 9.3 Let $(V, \langle ., . \rangle)$ be a Hilbert space of finite dimension. Show that every inner product on V can be written in the form $(v, w) = \langle Sv, Sw \rangle$ for some matrix $S \in \mathrm{GL}(V)$.

Exercise 9.4 Show that in $\mathrm{Mat}_n(\mathbb{C})$ every Cauchy sequence with respect to $\|.\|_1$ or $\|.\|_2$ converges.

(Hint: Show in either case that for a given Cauchy sequence all the entries are Cauchy sequences in $\mathbb{C}$.)

Exercise 9.5 Let $\rho : G \to GL(V)$ and $\tau : G \to GL(W)$ denote two finite-dimensional representations. Let $V \otimes W$ denote the tensor product of V and W. Show that

$$\rho \otimes \tau : \quad G \quad \to \quad GL(V \otimes W),$$
$$g \quad \mapsto \quad \rho(g) \otimes \tau(g),$$

defines a representation of G.

Exercise 9.6 Show that the commutator bracket in $Mat_n(\mathbb{C})$ given by

$$[X, Y] = XY - YX$$

satisfies $[X, Y] = -[Y, X]$ and

$$[X, [Y, Z]] + [Y, [Z, X]] + [Z, [X, Y]] = 0$$

(Jacobi identity).

Exercise 9.7 Show that for $A \in Mat_n(\mathbb{C})$ the function $f : \mathbb{R} \to Mat_n(\mathbb{C})$ given by

$$f(t) = \exp(tA)$$

is the unique solution of the matrix-valued differential equation

$$f'(t) = Af(t)$$

with $f(0) = 1$ and $f(t)A = Af(t)$.

Exercise 9.8 Let $A = \begin{pmatrix} 0 & 1 \\ 0 & 0 \end{pmatrix}$ and $B = \begin{pmatrix} 0 & 0 \\ 1 & 0 \end{pmatrix}$. Show that

$$\exp(A + B) \neq \exp(A)\exp(B).$$

Exercise 9.9 Show that for $A \in Mat_n(\mathbb{C})$ with $\|A - 1\|_1 < 1$ the series

$$\log(A) = -\sum_{n=1}^{\infty} \frac{(1 - A)^n}{n}$$

converges, and that for such A we have

$$\exp(\log(A)) = A.$$

(Hint: Prove the second claim for diagonal matrices first.)

Exercise 9.10 Show that if the metrizable group G is path connected, then it has no open subgroups other than itself. (Hint: Suppose that H is an open subgroup and assume that there is $x \in G \smallsetminus H$. Choose a path γ with $\gamma(0) = 1 \in G$ and $\gamma(1) = x$. Let t_0 be the infimum of the t with $\gamma(t) \in G \smallsetminus H$. Use the fact that H and $G \smallsetminus H$ are both closed (Exercise 6.22) to show that $\gamma(t_0)$ belongs to both of them, which is a contradiction.)

Exercise 9.11 Let $f(z) = \sum_{n=0}^{\infty} a_n z^n$ be a power series that converges for every $z \in \mathbb{C}$. Show that for every $A \in \mathrm{Mat}_n(\mathbb{C})$ the series

$$f(A) = \sum_{n=0}^{\infty} a_n A^n$$

converges to a matrix $f(A) \in \mathrm{Mat}_n(\mathbb{C})$. Show that the eigenvalues of $f(A)$ are all of the form $f(\lambda)$ for an eigenvalue λ of A.

Exercise 9.12 Show that the group $\mathrm{GL}_n(\mathbb{R})$ of real invertible $n \times n$ matrices is not path connected.

Exercise 9.13 Show that the group

$$\mathrm{SO}(2) = \left\{ \left(\begin{array}{cc} a & b \\ -b & a \end{array} \right) \middle| a, b \in \mathbb{R}, \ a^2 + b^2 = 1 \right\}$$

is isomorphic to $\mathbb{T}$.

Exercise 9.14 Let G be a closed subgroup of $\mathrm{GL}_n(\mathbb{C})$. Let $f : \mathrm{GL}_n(\mathbb{C}) \to [0, \infty)$ be a smooth function with compact support. Then $f|_G$ has compact support on G. Let $\pi : G \to \mathrm{GL}(V)$ be a finite-dimensional representation of G. Choose a Haar integral on G. For $v \in V$ let

$$\pi(f)v = \int_G f(x)\pi(x)v \, dx.$$

(a) Show that if f_n is a sequence of smooth functions as above such that $\int_G f_n = 1$ and such that the support of f_n shrinks to $\{e\}$ as n tends to infinity, then $\pi(f_n)v$ tends to v for every $v \in V$. Deduce that there is an f as above such that $\pi(f)V = V$.

(b) Show that for every $v \in V$ and every $X \in \mathrm{Lie}(G)$ the map $t \mapsto \pi(\exp(tX))v$ is smooth.

   (Hint: Write $v = \pi(f)w$ for some $w \in V$.)

Exercise 9.15 Let G be an arbitrary locally compact group. Choose a Haar measure and define the Hilbert space $L^2(G)$ as the completion of $C_c(G)$. For $\varphi \in C_c(G)$ and $x, y \in G$ define

$$L(y)\varphi(x) \stackrel{\mathrm{def}}{=} \varphi(y^{-1}x).$$

Show that $L(y)$ is unitary and extends to a unitary representation of G on $L^2(G)$, called the *left regular representation*. If G is unimodular, show that

$$R(y_1, y_2)\varphi(x) \stackrel{\mathrm{def}}{=} \varphi(y_1^{-1}xy_2)$$

defines a unitary representation of $G \times G$ on $L^2(\mathbb{R})$, called the *regular representation* of G.

Chapter 10

The Representations of SU(2)

For non-abelian groups, irreducible unitary representations play the part that characters play for abelian groups. Therefore, an obvious question is whether these representations can be classified. In the case of a compact connected matrix group we already have all that it takes to solve this problem.

In this chapter we are going to find all finite-dimensional irreducible representations of the group

$$
\begin{aligned}
\mathrm{SU}(2) &= \{A \in \mathrm{Mat}_2(\mathbb{C}) | A^*A = \mathbf{1},\ \det(A) = 1\} \\
&= \left\{ \begin{pmatrix} a & b \\ -\bar{b} & \bar{a} \end{pmatrix} \middle| a, b \in \mathbb{C},\ |a|^2 + |b|^2 = 1 \right\}.
\end{aligned}
$$

The second presentation shows that the group SU(2) is path connected.

The following result will be useful later.

Lemma 10.0.1 *Let K be a compact metrizable group and let ρ be a representation on a finite-dimensional Hilbert space $(V, \langle ., . \rangle)$. Then there is $S \in \mathrm{GL}(V)$ such that the representation $S\rho S^{-1}$ is unitary.*

Proof: Suppose we can show that there is a second inner product $(., .)$ on V such that ρ is unitary with respect to $(., .)$. Since every inner product on V is of the form $(v, w) = \langle Sv, Sw \rangle$ for some $S \in$

141

$GL(V)$, (see Exercise 10.1) it then follows that $S\rho S^{-1}$ is unitary. So
it remains to show that such an inner product exists. For $v, w \in V$,
let

$$(v, w) = \int_K \langle \rho(k^{-1})v, \rho(k^{-1})w \rangle \, dk.$$

It is easy to see that $(., .)$ is indeed an inner product. Moreover, the
representation ρ is unitary with respect to $(., .)$, since for $k_0 \in K$ and
$v, w \in V$ we have

$$
\begin{aligned}
(\rho(k_0)v, \rho(k_0)w) &= \int_K \langle \rho(k^{-1})\rho(k_0)v, \rho(k^{-1})\rho(k_0)w \rangle \, dk \\
&= \int_K \langle \rho((k_0^{-1}k)^{-1})v, \rho((k_0^{-1}k)^{-1})w \rangle \, dk \\
&= \int_K \langle \rho(k^{-1})v, \rho(k^{-1})w \rangle \, dk \\
&= (v, w),
\end{aligned}
$$

which implies the lemma. $\square$

10.1   The Lie Algebra

The Lie algebra of SU(2) is the algebra of all skew-adjoint trace zero
matrices; i.e., the Lie algebra is

$$su(2) = \{X \in \mathrm{Mat}_2(\mathbb{C}) | X^* = -X, \ \mathrm{tr}(X) = 0\}.$$

We fix a standard basis of $su(2)$ consisting of

$$X_1 = \frac{1}{2}\begin{pmatrix} i & \\ & -i \end{pmatrix}, \quad X_2 = \frac{1}{2}\begin{pmatrix} & 1 \\ -1 & \end{pmatrix}, \quad X_3 = \frac{1}{2}\begin{pmatrix} & i \\ i & \end{pmatrix}.$$

The following relations are easily verified by direct computation:

$$[X_1, X_2] = X_3, \quad [X_2, X_3] = X_1, \quad [X_3, X_1] = X_2.$$

Let $\pi : su(2) \to \mathrm{End}(V)$ be a finite-dimensional $*$-representation of
the Lie algebra $su(2)$. Let $L_j = \pi(X_j) \in \mathrm{End}(V)$ for $j = 1, 2, 3$. It
then follows that $[L_1, L_2] = L_3$, $[L_2, L_3] = L_1$, $[L_3, L_1] = L_2$, and
for $j = 1, 2, 3$,

$$L_j^* = \pi(X_j)^* = \pi(X_j^*) = \pi(-X_j) = -L_j,$$

so each L_j is a skew-adjoint operator. In particular, it follows that L_j is diagonalizable (see Exercise 9.1). For every $\mu \in \mathbb{C}$ let

$$V_\mu = \{v \in V | L_1 v = i\mu v\}.$$

Then the space V decomposes into a direct sum of eigenspaces:

$$V = \bigoplus_{i\mu \in \operatorname{spec}(L_1)} V_\mu,$$

where $\operatorname{spec}(L_1)$ denotes the spectrum of L_1, i.e., in this case the set of eigenvalues, which is a subset of $i\mathbb{R}$. Let

$$L_+ = L_2 - iL_3, \quad L_- = L_2 + iL_3.$$

A computation shows that

$$[L_1, L_\pm] = \pm iL_\pm.$$

Proposition 10.1.1 *The operator $L_\pm$ maps V_μ to $V_{\mu\pm1}$. In particular, if $i\mu \in \operatorname{spec}(L_1)$, then either L_+ is zero on V_μ or $i(\mu + 1) \in \operatorname{spec}(L_1)$.*

Proof: Let $v \in V_\mu$; then

$$L_1(L_+v) = L_+L_1v + iL_+v = i(\mu + 1)L_+v.$$

This implies $L_+V_\mu \subset V_{\mu+1}$. Similarly, it follows that also $L_-V_\mu \subset V_{\mu-1}$. $\square$

Let $C = L_1^2 + L_2^2 + L_3^2$; then a computation shows that

$$CL_j = L_jC \quad \text{for } j = 1, 2, 3.$$

Lemma 10.1.2 *If π is irreducible, then there is a $\lambda \in \mathbb{C}$ such that $C = \lambda\operatorname{Id}$.*

Proof: Let λ be an eigenvalue of C. Then, since the L_j commute with C, they leave the corresponding eigenspace invariant, so this is an invariant subspace. If π is irreducible, then this subspace must be all of V. $\square$

Proposition 10.1.3 *Let π be an irreducible representation of $su(2)$ on V. Then the spectrum of L_1 is a sequence*

$$\{i\mu_0, i(\mu_0 + 1), \ldots, i(\mu_0 + k) = i\mu_1\}$$

with

$$L_+ : V_{\mu_0+j} \to V_{\mu_0+j+1}$$

an isomorphism for $0 \le j \le k - 1$, and

$$L_- : V_{\mu_1-j} \to V_{\mu_1-j-1}$$

an isomorphism for $0 \le j \le k - 1$. The situation is depicted as follows:

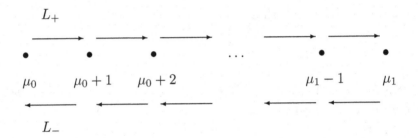

Moreover, the spaces V_{μ_0+j} are one-dimensional for $j = 0, 1, \ldots, k$, and so $\dim V = k + 1$. Finally, $\mu_0 = -k/2$, and so $\mu_1 = k/2$. In particular, it follows that every two finite dimensional irreducible representations of the Lie algebra of $SU(2)$ are isomorphic if they have the same dimension.

Proof: By Lemma 10.0.1 we may assume the representation to be unitary. We compute

$$
\begin{aligned}
L_-L_+ &= (L_2 + iL_3)(L_2 - iL_3) = L_2^2 + L_3^2 + i[L_3, L_2] \\
&= C - L_1^2 - iL_1 = \lambda - L_1^2 - iL_1,
\end{aligned}
$$

and

$$L_+L_- = C - L_1^2 + iL_1 = \lambda - L_1^2 + iL_1.$$

So on V_μ we have

$$
\begin{aligned}
L_+L_- &= \lambda + \mu(\mu - 1), \\
L_-L_+ &= \lambda + \mu(\mu + 1).
\end{aligned}
$$

Since L_2 and L_3 are skew-adjoint, we have

$$L_+^* = (L_2 - iL_3)^* = -L_-.$$

This implies that the operators L_+L_- and L_-L_+ are self-adjoint.

Lemma 10.1.4 Let V be a finite-dimensional Hilbert space and let A be a linear operator on V; then we have

$$\ker A = \ker A^*A.$$

Proof: For $v \in V$ we have

$$
\begin{aligned}
v \in \ker(A) &\Leftrightarrow Av = 0 \\
&\Leftrightarrow \langle Av, Aw \rangle = 0 \ \forall w \in V \\
&\Leftrightarrow \langle A^*Av, w \rangle = 0 \ \forall w \in V \\
&\Leftrightarrow A^*Av = 0 \\
&\Leftrightarrow v \in \ker(A^*A).
\end{aligned}
$$

The lemma is proven. $\qquad\square$

The lemma implies

$$
\begin{aligned}
\ker L_- &= \ker L_+L_-, \\
\ker L_+ &= \ker L_-L_+.
\end{aligned}
$$

Now let $\mu_0, \mu_0 + 1, \ldots, \mu_0 + k = \mu_1$ be a sequence of maximal length, with $V_{\mu_0+j} \neq 0$ for $j = 0, \ldots, k$. Then it follows that $L_+V_{\mu_0+k} = 0$, and therefore

$$0 = \lambda + \mu_1(\mu_1 + 1) = \lambda + \mu_0(\mu_0 - 1),$$

or

$$\mu_0(\mu_0 - 1) = -\lambda = \mu_1(\mu_1 + 1).$$

This implies that

$$\mu_0(\mu_0 - 1) = (\mu_0 + k)(\mu_0 + k + 1),$$

or

$$-\mu_0 = \mu_0(2k + 1) + k(k + 1),$$

which implies $\mu_0 = -k/2$. Now let $v \in V_{\mu_0}$ and suppose $v \neq 0$. Let $V'_{\mu_0+j} = \mathbb{C}L_+^j v$. The space $V' = V'_{\mu_0} \oplus \cdots \oplus V'_{\mu_0+k}$ is preserved by L_1, L_2, and L_3, and hence by $\mathrm{Lie}(G)$, so it is an invariant subspace. Since π is irreducible, it follows that $V = V'$, and so in particular, $V'_{\mu_0+j} = V_{\mu_0+j}$ for all j, and so the spaces V_{μ_0+j} are all one-dimensional. $\qquad\square$

10.2 The Representations

We shall now use our knowledge about the representations of the Lie algebra to classify the representations of the group SU(2).

Lemma 10.2.1 *Given a finite-dimensional representation ρ of the group* SU(2), *suppose the subgroup T of diagonal matrices in* SU(2) *acts by the characters* $\chi_{-k}, \chi_{-k+2}, \ldots, \chi_k$, *where*

$$\chi_j \begin{pmatrix} \varepsilon & 0 \\ 0 & \bar{\varepsilon} \end{pmatrix} = \varepsilon^j.$$

Assume that the eigenspaces for the χ_j are all one-dimensional. Then ρ is irreducible.

Proof: This follows from the proposition, since ρ must already be irreducible under the Lie algebra. □

Theorem 10.2.2 *For each $k \in \{0, 1, 2, \ldots\}$ there is exactly one irreducible representation of* SU(2) *of dimension $k + 1$.*

Proof: Let k be in the set $\{0, 1, 2, \ldots\}$ and let V_k be the set of all homogeneous polynomials of degree k in the two variables x, y. Then

$$V_k = \mathbb{C}x^k \oplus \mathbb{C}x^{k-1}y \oplus \cdots \oplus \mathbb{C}y^k.$$

Let $\rho_k : \mathrm{SU}(2) \to \mathrm{GL}(V_k)$ be defined by

$$\rho_k(A)f(x, y) = f((x, y)A),$$

i.e.,

$$\rho_k\left(\begin{pmatrix} a & b \\ c & d \end{pmatrix}\right) = f(ax + cy, bx + dy).$$

By Lemma 10.2.1 it follows that ρ_k is irreducible. By Proposition 10.1.3 it follows that ρ_k is the unique irreducible representation of $su(2)$ of dimension $k + 1$, and finally, by Lemma 9.3.6, ρ_k is the unique SU(2)-representation of dimension $k + 1$. □

10.3 Exercises

Exercise 10.1 Let $(.,.)$ denote an inner product on the vector space $\mathbb{C}^n$. Show that there exists a matrix $S \in \mathrm{GL}_n(\mathbb{C})$ such that for all $v, w \in \mathbb{C}^n$,

$$(v, w) = \langle Sv, Sw \rangle,$$

where $\langle v, w \rangle = v^t \overline{w}$ is the standard inner product on $\mathbb{C}^n$.

Exercise 10.2 Show that for an LCA group A every irreducible finite-dimensional unitary representation is one-dimensional.

Exercise 10.3 Show that $\mathrm{U}(1) \cong \mathbb{T}$, and determine the set $\widehat{\mathrm{U}(1)}$.

Exercise 10.4 Show that $\mathrm{U}(2) \cong (\mathrm{U}(1) \times \mathrm{SU}(2))/\{\pm 1\}$, and determine $\widehat{\mathrm{U}(2)}$.

Exercise 10.5 Let (τ, V) and (ρ, W) be finite-dimensional representations of an LC group G. On the tensor product $V \otimes W$ define a representation $\tau \otimes \rho$ by $\tau \otimes \rho(g) = \tau(g) \otimes \rho(g)$. For $G = \mathrm{SU}(2)$ show that

$$\rho_k \otimes \rho_l \cong \rho_{k+l} \oplus \rho_{k+l-2} \oplus \cdots \oplus \rho_{|k-l|}.$$

Chapter 11

The Peter-Weyl Theorem

The Peter-Weyl theorem generalizes the completeness of the Fourier series, and so it is Plancherel's theorem for compact groups. It states that for a compact group K the matrix coefficients of the finite-dimensional irreducible unitary representations give an orthonormal basis of $L^2(K)$. We will prove it here only for matrix groups.

11.1 Decomposition of Representations

Lemma 11.1.1 Let (π, V_π) be a finite-dimensional unitary representation of the LC group G. Then π splits into a direct sum of irreducible representations.

Proof: This is proven by induction on the dimension of V_π. If V_π is one-dimensional, then the representation is clearly irreducible and we are done. Now suppose the claim is proven for all spaces of dimension lower than the dimension of V_π. Then either V_π is irreducible, in which case we are finished, or it has a proper subrepresentation W. But then the orthogonal space $W^\perp = \{v \in V_\pi | \langle v, w \rangle = 0 \ \forall w \in W\}$ is also G-invariant, as follows from the unitarity of π. So then V_π is the direct sum of the subrepresentation spaces W and $W^\perp$, which are both of smaller dimension, and hence decompose into irreducibles, and so then does V_π. $\qquad\square$

11.2   The Representation on $\mathrm{Hom}(V_\gamma, V_\tau)$

Let K be a compact matrix group, i.e., a compact subgroup of $GL_n(\mathbb{C})$. Let τ and γ be irreducible finite-dimensional representations of K. Let H be the space of all linear maps from V_γ to V_τ. On this space we define a new representation η of K by

$$\eta(k)T = \tau(k)T\gamma(k^{-1}).$$

Let $\mathrm{Hom}_K(V_\gamma, V_\tau)$ be the space of K-homomorphisms, i.e., the space of all linear maps $T: V_\gamma \to V_\tau$ such that

$$T\gamma(k) = \tau(k)T$$

for every $k \in K$.

For every representation (σ, V_σ) of K let

$$V_\sigma^K = \{v \in V_\sigma | \sigma(k)v = v \ \forall k \in K\};$$

i.e., V_σ^K is the space of K-fixed vectors.

Lemma 11.2.1 *The space*

$$H^K = \mathrm{Hom}_{\mathbb{C}}(V_\gamma, V_\tau)^K = \mathrm{Hom}_K(V_\gamma, V_\tau)$$

is at most one-dimensional.

Proof: Let $T \in H^K$ and assume that $T \neq 0$. Then the kernel $\ker(T)$ is an invariant subspace of V_γ, since $v \in \ker(T)$ implies for every $k \in K$ that $T(\gamma(k)v) = \tau(k)Tv = 0$. Therefore, $\gamma(k)v$ again lies in $\ker(T)$, which is thus invariant. Since γ is irreducible, it follows that if $T \neq 0$, then T is injective. Likewise, the image of T is an invariant subspace, and since τ is irreducible, too, it follows that T is surjective, and hence is an isomorphism.

Finally, let $T, S \in H^K$, and assume that both are nonzero. Then both are invertible, and $S^{-1}T$ is in $\mathrm{Hom}_K(V_\gamma, V_\gamma)$. Let λ be an eigenvalue of $S^{-1}T$. Then the corresponding eigenspace $\mathrm{Eig}(\lambda)$ is invariant, so by the irreducibility it follows that $\mathrm{Eig}(\lambda) = V_\gamma$ or $S^{-1}T = \lambda\mathrm{Id}$. Hence $T = \lambda S$, so the dimension of the space H^K is at most one. $\square$

Two unitary representations γ and τ of K are called *isomorphic* if there is a unitary map $T : V_\gamma \to V_\tau$ satisfying $T\gamma(k) = \tau(k)T$ for every $k \in K$. Let $\hat{K}_{\text{fin}}$ be the set of all isomorphism classes of finite-dimensional irreducible unitary representations of K.

Lemma 11.2.2 *Two finite-dimensional irreducible representations γ, τ of K are isomorphic if and only if*

$$\text{Hom}_K(V_\gamma, V_\tau) \neq 0,$$

regardless of the inner products.

Proof: If the two representations γ and τ are isomorphic, there exists a K-homomorphism between them, so one direction is clear. Conversely, suppose $T \neq 0$ lies in $\text{Hom}_K(V_\gamma, V_\tau)$. We will show that there is $\lambda \in \mathbb{C}$ such that λT is unitary, i.e., $(\lambda T)^*(\lambda T) = \text{Id}$, or $|\lambda|^2 T^* T = \text{Id}$. Now, $T^* T$ is a K-homomorphism and so is the identity. Thus by Lemma 11.2.1 there is $c \in \mathbb{C}$ such that $T^* T = c\,\text{Id}$. The operator $T^* T$ is positive self-adjoint, so $c > 0$. Therefore, there is $\lambda \in \mathbb{C}$ such that $c = 1/|\lambda|^2$. $\qquad\square$

11.3 The Peter-Weyl Theorem

For each class in $\hat{K}_{\text{fin}}$ choose a fixed representative (τ, V_τ). Choose an orthonormal basis $e_1, \dots, e_n$ of V_τ and let

$$\tau_{i,j}(k) = \langle \tau(k)e_i, e_j \rangle.$$

The map $\tau_{i,j} : K \to \mathbb{C}$ is called the (i,j)th *matrix coefficient* of τ.

Theorem 11.3.1 *(Peter-Weyl) Let $\tau \neq \gamma$ in $\hat{K}_{\text{fin}}$. Then*

$$\int_K \tau_{i,j}(k)\overline{\gamma_{r,s}(k)}dk = 0$$

for all indices i, j, r, s. Further,

$$\int_K \tau_{i,j}(k)\overline{\tau_{r,s}(k)}dk = 0$$

unless $i = r$ and $j = s$. In addition,

$$\int_K \tau_{i,j}(k)\overline{\tau_{i,j}(k)}dk = \frac{1}{\dim(V_\tau)}.$$

The family $(\sqrt{\dim(V_\tau)}\tau_{i,j})_{\tau,i,j}$ forms an orthonormal basis of the Hilbert space $L^2(K)$.

Proof: Assume first $\tau \neq \gamma$. Then for every $T \in \mathrm{Hom}_{\mathbb{C}}(V_\gamma, V_\tau)$ we have that

$$S = \int_K \tau(k)T\gamma(k^{-1})dk$$

satisfies $\tau(k)S = S\gamma(k)$ for every $k \in K$, and therefore $S = 0$. Let $e_1, \ldots, e_n$ be the given basis of V_γ and let $f_1, \ldots, f_m$ be the given basis of V_τ. Let $T \in \mathrm{Hom}_{\mathbb{C}}(V_\gamma, V_\tau)$ be given by the matrix $E_{i,j}$ with a one at position (i, j) and zeros everywhere else. Then $\tau(k)T\gamma(k^{-1})$ is given by the following matrices $\gamma(k^{-1})$:

$$\begin{pmatrix} \tau_{1,1} & \cdots & \tau_{1,m} \\ \vdots & & \vdots \\ \tau_{m,1} & \cdots & \tau_{m,m} \end{pmatrix} \begin{pmatrix} & & \\ & 1 & \\ & & \end{pmatrix} \begin{pmatrix} \bar{\gamma}_{1,1} & \cdots & \bar{\gamma}_{n,1} \\ \vdots & & \vdots \\ \bar{\gamma}_{1,n} & \cdots & \bar{\gamma}_{n,n} \end{pmatrix}$$

$$= \begin{pmatrix} \tau_{1,i} \\ \vdots \\ \tau_{m,i} \end{pmatrix} \begin{pmatrix} \bar{\gamma}_{1,1} & \cdots & \bar{\gamma}_{n,1} \\ \vdots & & \vdots \\ \bar{\gamma}_{1,n} & \cdots & \bar{\gamma}_{n,n} \end{pmatrix}$$

$$= \begin{pmatrix} \tau_{1,i}\bar{\gamma}_{1,j} & \cdots & \tau_{1,i}\bar{\gamma}_{n,j} \\ \vdots & & \vdots \\ \tau_{m,i}\bar{\gamma}_{1,j} & \cdots & \tau_{m,i}\bar{\gamma}_{n,j} \end{pmatrix}.$$

By varying i and j we get the first part of the theorem. For $\gamma = \tau$ we have

$$\int_K \tau(k)T\gamma(k^{-1})dk = \lambda\mathrm{Id}$$

for some $\lambda \in \mathbb{C}$, which implies the second assertion of the theorem. To see that $\int_K |\tau_{i,j}(k)|^2 dk = 1/\dim(V_\tau)$, let $T = \mathrm{Id}$ and recall that

$$\tau(k)\tau(k)^* = \mathrm{Id}.$$

This implies

$$\sum_r \tau_{i,r}\overline{\tau_{j,r}(k)} = \delta_{i,j},$$

so that for $i = j$ we have

$$\sum_r |\tau_{i,r}(k)|^2 = 1,$$

so

$$\sum_r \int_K |\tau_{i,r}(k)|^2 dk \;=\; 1.$$

Now the fact that $\int_K \tau(k) T \tau(k^{-1}) dk = \lambda \mathrm{Id}$ applied in the case $T = E_{i,i}$ implies that

$$\int_K |\tau_{r,i}(k)|^2 dk \;=\; \int_K |\tau_{r',i}(k)|^2 dk$$

for every two r, r'. Finally, by Exercise 8.6

$$\int_K |\tau_{r,i}(k)|^2 dk \;=\; \int_K |\tau_{r,i}(k^{-1})|^2 dk \;=\; \int_K |\tau_{i,r}(k)|^2 dk,$$

and so we get

$$\int_K |\tau_{i,r}(k)|^2 dk \;=\; \int_K |\tau_{i,r'}(k)|^2 dk$$

for every two r, r', which implies the theorem, save for the completeness of the system $(\sqrt{\dim(V_\tau)}\tau_{i,j})$.

To establish the completeness we will use the following weak version of the Stone-Weierstrass theorem. Let X be a compact metrizable space. On the space $C(X)$ of all continuous complex-valued functions on X we define a norm by

$$\|f\|_\infty \;=\; \sup_{x \in X} |f(x)|.$$

This gives a metric defined by $d(f, g) = \|f - g\|_\infty$. The space $C(X)$ is a $\mathbb{C}$-algebra by pointwise multiplication.

Lemma 11.3.2 *(Stone-Weierstrass theorem) Let X be a compact metrizable space and let A be a subalgebra of $C(X)$ such that*

- *A is closed under complex conjugation, i.e., $f \in A$ implies $\bar{f} \in A$,*

- *A separates points, i.e., for every two distinct points $x, y \in X$ there exists $f \in A$ such that $f(x) \neq f(y)$, and*

- *for every $x \in X$ there is $f \in A$ such that $f(x) \neq 0$.*

Then A is dense in $C(X)$.

Proof: See [21].   □

Now the Peter-Weyl theorem is easily deduced. Let A be the linear span of all matrix coefficients $\tau_{i,j} \in C(K)$. Then A is closed under complex conjugation, since the conjugate $\bar{\tau}$ of τ again is a representation of K. Next, A is closed under multiplication, since as we have seen, the product of the coefficients $\tau_{i,j}$ and $\gamma_{r,s}$ occurs as a coefficient in the representation on $\mathrm{Hom}_{\mathbb{C}}(V_{\bar{\gamma}}, V_{\tau})$. Finally, A separates points, since K is a matrix group, so it has an injective representation and hence by Lemma 10.0.1 also an injective unitary representation; this also means that the last condition in the statement of the theorem is fulfilled. The Stone-Weierstrass theorem applies to show that A is dense in $C(K)$, and hence it is dense in $L^2(K)$, since $C(K)$ is. Therefore the system is complete.   □

11.4   A Reformulation

Let K be a compact matrix group. The group $K \times K$ acts on the space $C(K)$ of all continuous functions on K by

$$(k_1, k_2).f(k) \;=\; f(k_1^{-1} k k_2).$$

Fix a Haar integral on K. Then this action is unitary, and hence extends to the L^2-completion $L^2(K)$, which then becomes a $K \times K$ unitary representation space. The matrix coefficients of representations in $\hat{K}_{\mathrm{fin}}$ give elements in $L^2(K)$, and so the Peter-Weyl theorem can be restated as follows:

Theorem 11.4.1 *The matrix coefficients define a $K \times K$ isomorphism*

$$L^2(K) \;\cong\; \hat{\bigoplus_{\tau \in \hat{K}_{\mathrm{fin}}}} \mathrm{End}(V_\tau),$$

where $\hat{\oplus}$ means the Hilbert space completion of the algebraic direct sum.

For $f \in L^1(K) \subset L^2(K)$ the image of f under this isomorphism is given by

$$\sum_{\tau \in \hat{K}_{\mathrm{fin}}} \tau(f),$$

where

$$\tau(f) = \int_K f(k)\tau(k)dk.$$

For a given irreducible unitary representation $\tau : K \to \mathrm{GL}(V)$ the matrix coefficients give an embedding $V \hookrightarrow L^2(K)$, so by the Peter-Weyl theorem we conclude that V must decompose into a direct sum of finite-dimensional representations, which by irreducibility implies that τ itself is finite-dimensional. This leads to the following result.

Theorem 11.4.2 *Every irreducible unitary representation of a compact matrix group K is finite-dimensional, and thus $\hat{K}_{\mathrm{fin}}$ coincides with $\hat{K}$, the set of all irreducible unitary representations of K modulo isomorphism.*

For noncompact groups the theorem does not hold, as the example $\mathrm{SL}_2(\mathbb{R})$ shows [14]. But a version of Plancherel's theorem still holds for a general matrix group G that is unimodular. It turns out that the $G \times G$ representation space $L^2(G)$ is then isomorphic not to a direct sum over the set $\hat{G}$ of all isomorphism classes of irreducible unitary representations, but rather to a direct Hilbert integral; see [3] for details.

11.5 Exercises

Exercise 11.1 Let (ρ, V) be a finite-dimensional irreducible representation of the compact matrix group K. Show that any two K-invariant inner products on V differ by a constant factor.

Exercise 11.2 Let K be a compact matrix group. For $f, g \in C(K)$ define their convolution by

$$f * g(x) = \int_K f(y)g(y^{-1}x)dy.$$

Show that $f * g \in C(K)$ and show that K is abelian if and only if the convolution algebra $C(K)$ is commutative, i.e., $f * g = g * f$ holds for every $f, g \in C(K)$.

Exercise 11.3 Show that a compact matrix group K is abelian if and only if every irreducible unitary representation of K is one-dimensional.

Exercise 11.4 Let K be a compact matrix group. For $\tau \in \hat{K}$ the function $\chi_\tau : K \to \mathbb{C}$ defined by

$$\chi_\tau(k) \;=\; \operatorname{tr} \tau(k)$$

is called the character of τ. Show that for $\tau\eta \in \hat{K}$,

$$\langle \chi_\tau, \chi_\eta \rangle \;=\; \begin{cases} 1 & \text{if } \tau = \eta, \\ 0 & \text{otherwise.} \end{cases}$$

Exercise 11.5 Let K be a compact matrix group, and let $H \subset K$ be a closed subgroup. Show that the unitary representation of K given by left translation on the space $L^2(K/H)$ is isomorphic to

$$\bigoplus_{\tau \in \hat{K}} \dim(V_\tau^H)\tau,$$

where V_τ^H is the subspace of V_τ of H-invariants, i.e.,

$$V_\tau^H \;=\; \{v \in V_\tau \mid \tau(h)v = v \ \forall h \in H\}.$$

Chapter 12

The Heisenberg Group

In this chapter we give an example of a group that is neither abelian nor compact. The general phenomenon in the harmonic analysis of such groups G is that the regular representation on $L^2(G)$ can be decomposed into a direct integral over the unitary dual $\hat{G}$. This is the most general form of a Plancherel theorem. We will not go into the details of direct integrals here, but we will treat the example of the Heisenberg group in some detail.

When we speak of a Hilbert space we usually mean an infinite-dimensional separable Hilbert space.

12.1 Definition

The Heisenberg group $\mathcal{H}$ is defined to be the group of real upper triangular 3×3 matrices with ones on the diagonal:

$$\mathcal{H} \stackrel{\text{def}}{=} \left\{ \begin{pmatrix} 1 & x & z \\ & 1 & y \\ & & 1 \end{pmatrix} \middle| x, y, z \in \mathbb{R} \right\}.$$

It can also be identified with $\mathbb{R}^3$, where the group law is

$$(a, b, c)(x, y, z) \stackrel{\text{def}}{=} (a + x, b + y, c + z + ay).$$

The inverse of (a, b, c) is

$$(a, b, c)^{-1} = (-a, -b, ab - c).$$

157

The center of $\mathcal{H}$ is $Z(\mathcal{H}) = \{(0, 0, z) \mid z \in \mathbb{R}\}$, and we have

$$\mathcal{H}/Z(\mathcal{H}) \cong \mathbb{R}^2.$$

Lemma 12.1.1 *A Haar integral on $\mathcal{H}$ is given by*

$$\int_{\mathcal{H}} f(h)\, dh \stackrel{\text{def}}{=} \int_{\mathbb{R}} \int_{\mathbb{R}} \int_{\mathbb{R}} f(a, b, c)\, da\, db\, dc, \qquad f \in C_c(\mathcal{H}).$$

This Haar integral is left- and right-invariant, so $\mathcal{H}$ is unimodular.

We will use this Haar measure on $\mathcal{H}$ for all computations in the sequel.

Proof: Let $(\alpha, \beta, \gamma) \in \mathcal{H}$. For $f \in C_c(\mathcal{H})$ we compute

$$\int_{\mathcal{H}} f((\alpha, \beta, \gamma)h)\, dh \;=\; \int_{\mathbb{R}^3} f((\alpha, \beta, \gamma)(a, b, c))\, da\, db\, dc$$

$$= \int_{\mathbb{R}^3} f(a + \alpha, b + \beta, c + \gamma + \alpha b)\, da\, db\, dc$$

$$= \int_{\mathbb{R}^2} \left(\int_{\mathbb{R}} f(a + \alpha, b + \beta, c + \gamma + \alpha b)\, dc \right)\, da\, db$$

$$= \int_{\mathbb{R}^2} \left(\int_{\mathbb{R}} f(a + \alpha, b + \beta, c)\, dc \right)\, da\, db$$

$$= \int_{\mathbb{R}^3} f(a, b, c)\, da\, db\, dc.$$

The right translation is dealt with in a similar fashion. $\square$

12.2   The Unitary Dual

For a locally compact group G, two unitary representations (π, V_π) and (η, V_η) are called *isomorphic* or *unitarily equivalent* if there exists a unitary operator $T \colon V_\pi \to V_\eta$ with

$$T\pi(g) \;=\; \eta(g)\, T$$

for every $g \in G$. Since this actually means that $\eta = T\pi T^{-1}$, it follows that π and η are indistinguishable as representations.

Isomorphism is an equivalence relation on the class of unitary representations. The set of equivalence classes

$$\hat{G} \stackrel{\text{def}}{=} \{\pi \text{ irreducible unitary }\}/\text{isomorphy}$$

is called the *unitary dual* of G. It is the substitute for the dual group in the case of abelian groups. We will often write $\pi \in \hat{G}$ when we actually mean the class of the representation π. If G is abelian, this notation appears to be ambiguous, since the dual group was already named $\hat{G}$, but in this case the unitary dual can be identified with the dual group (see Exercise 12.1).

In this section we are going to describe the unitary dual $\hat{\mathcal{H}}$ of the Heisenberg group $\mathcal{H}$.

Let $\hat{\mathcal{H}}_0$ denote the subset of $\hat{\mathcal{H}}$ consisting of all classes $\pi \in \hat{\mathcal{H}}$ such that $\pi(h) = 1$ whenever h lies in the center $Z(\mathcal{H})$ of $\mathcal{H}$. Since $\mathcal{H}/Z(\mathcal{H}) \cong \mathbb{R}^2$, it follows that

$$\hat{\mathcal{H}}_0 = \widehat{\mathcal{H}/Z(\mathcal{H})} \cong \widehat{\mathbb{R}^2} \cong \hat{\mathbb{R}}^2,$$

and the latter can be identified with $\mathbb{R}^2$ in the following explicit way. Let $(a, b) \in \mathbb{R}^2$ and define a character

$$\begin{aligned} \chi_{a,b} \colon \quad \mathcal{H} &\to \mathbb{T}, \\ (x, y, z) &\mapsto e^{2\pi i(ax+by)}. \end{aligned}$$

The identification is given by $(a, b) \mapsto \chi_{a,b}$. In particular, it follows that all representations in $\hat{\mathcal{H}}_0$ are onedimensional. This observation indicates the importance of the behavior of the center under a representation.

Lemma 12.2.1 *Let (π, V_π) be an irreducible unitary representation of a locally compact group G. Let $Z(G) \subset G$ be the center of G. Then for every $z \in Z(G)$ the operator $\pi(z)$ on V_π is a multiple of the identity.*

Proof: Let $z \in Z(G)$. Then $\pi(z) \colon V_\pi \to V_\pi$ is a unitary operator, so its spectrum $\operatorname{spec} \pi(z)$ is contained in $\mathbb{T} = \{w \in \mathbb{C} \mid |w| = 1\}$. According to the spectral resolution for unitary operators [25], there are disjoint projections $F(\lambda)$ for $\lambda \in \mathbb{T}$ such that

$$\pi(z) = \int_{\mathbb{T}} \lambda \, dF(\lambda).$$

Each of the projections $F(\lambda)$ must commute with G, i.e., $\pi(g)F(\lambda) = F(\lambda)\pi(g)$ for every $g \in G$. Hence the image and kernel of $F(\lambda)$ are invariant subspaces. Since π is irreducible, it follows that $F(\lambda_0) = \mathrm{Id}$ for one λ_0, and $F(\lambda) = 0$ for $\lambda \ne \lambda_0$. This implies that $\pi(z)$ equals λ_0 times the identity. $\square$

As a consequence of the lemma, for each $\pi \in \hat{G}$ there is a character $\chi_\pi : Z(G) \to \mathbb{T}$ with $\pi(z) = \chi_\pi(z)\mathrm{Id}$ for every $z \in Z(G)$. This character χ_π is called the *central character* of the representation π.

For every character $\chi \ne 1$ of $Z(\mathcal{H})$ we will now construct an irreducible unitary representation of the Heisenberg group that has χ for its central character. We will start with a particular character, namely,

$$\chi_1(0,0,c) = e^{2\pi i c}.$$

As it will turn out, the group of unitary operators on $L^2(\mathbb{R})$ generated by

$$\varphi(x) \mapsto \varphi(x+a), \quad \varphi(x) \mapsto e^{2\pi i b x}\varphi(x),$$

where a and b vary in $\mathbb{R}$, is isomorphic to $\mathcal{H}$. With this in mind we let $(a,b,c) \in \mathcal{H}$ and define the operator $\pi_1(a,b,c)$ on $L^2(\mathbb{R})$ by

$$\pi_1(a,b,c)\varphi(x) \overset{\text{def}}{=} e^{2\pi i(bx+c)}\varphi(x+a).$$

To verify that π_1 is indeed a representation, one computes

$$
\begin{aligned}
\pi_1(a,b,c)\pi_1(\alpha,\beta,\gamma)\varphi(x) &= e^{2\pi i(bx+c)}\pi_1(\alpha,\beta,\gamma)\varphi(x+a)\\
&= e^{2\pi i(bx+c)}e^{2\pi i(\beta(x+a)+\gamma)}\varphi(x+a+\gamma)\\
&= e^{2\pi i((b+\beta)x+c+\gamma+a\beta)}\varphi(x+a+\alpha)\\
&= \pi_1(a+\alpha,b+\beta,c+\gamma+a\beta)\,\varphi(x)\\
&= \pi_1((a,b,c)(\alpha,\beta,\gamma))\varphi(x).
\end{aligned}
$$

It is immediate that the representation π_1 is unitary.

Lemma 12.2.2 π_1 *is irreducible.*

Proof: Let $V \subset L^2(\mathbb{R})$ be a closed invariant subspace with $V \ne 0$. If $\varphi \in V$, then $\pi(a,0,0)\varphi(x) = \varphi(x+a) \in V$, and so

$$\psi * \varphi(x) = \int_{\mathbb{R}} \psi(a)\varphi(x-a)\,da = \int_{\mathbb{R}} \psi(a+x)\varphi(-a)\,da$$

lies in V for every $\psi \in \mathcal{S}(\mathbb{R})$. Since these convolution products are smooth functions, we infer that V contains a nonzero smooth function φ_0. It follows that $\pi(0, b, 0)\varphi_0(x) = e^{-2\pi i b x}\varphi_0(x)$ lies in V, and so

$$\hat{\psi}(x)\varphi_0(x) = \int_{\mathbb{R}} \psi(b)e^{-2\pi i b x}\, dx \, \varphi_0(x)$$

lies in V for every $\psi \in \mathcal{S}(\mathbb{R})$. Choose some interval where φ_0 has no zeros, so that we get $C_c^\infty(I) \subset V$. By translation and summation we infer that $C_c^\infty(\mathbb{R}) \subset V$, and since $C_c^\infty(\mathbb{R})$ is dense in $L^2(\mathbb{R})$, we get $V = L^2(\mathbb{R})$, so π_1 is indeed irreducible. $\qquad \square$

Next we construct irreducible unitary representations for all other nontrivial characters. For this note that for $t \in \mathbb{R}^\times = \mathbb{R} \smallsetminus \{0\}$ the map

$$\theta_t(a, b, c) \overset{\text{def}}{=} (a, bt, ct)$$

is a continuous automorphism of $\mathcal{H}$, as is seen from

$$\begin{aligned}
\theta_t((a, b, c)(\alpha, \beta, \gamma)) &= \theta_t(a + \alpha, b + \beta, c + \gamma + a\beta) \\
&= (a + \alpha, (b + \beta)t, (c + \gamma + a\beta)t) \\
&= (a, bt, ct)(\alpha, \beta t, \gamma t) \\
&= \theta_t(a, b, c)\theta_t(\alpha, \beta, \gamma)
\end{aligned}$$

and $\theta_t^{-1} = \theta_{t^{-1}}$. Now set

$$\pi_t \overset{\text{def}}{=} \pi_1 \circ \theta_t, \qquad t \in \mathbb{R}, \ t \neq 0.$$

Since the image of π_t equals the image of π_1, it follows that π_t is irreducible. Let χ_t be its central character. We compute

$$\chi_t(0, 0, c) = \chi_1(\theta_t(0, 0, c)) = \chi_1(0, 0, ct) = e^{2\pi i c t}.$$

It can be shown [23] that up to isomorphy π_t is the only irreducible unitary representation of $\mathcal{H}$ with central character χ_t. Since this result is not important for the Plancherel Theorem, we will not give its proof. It yields, however, the following description of the unitary dual:

$$\hat{\mathcal{H}} = \hat{\mathbb{R}}^2 \cup \{\pi_t \mid t \in \mathbb{R}^\times\}.$$

12.3   Hilbert-Schmidt Operators

For a linear operator $T\colon H \to H$ on a Hilbert space H we define the *operator norm* by

$$\|T\| \overset{\text{def}}{=} \sup_{\|v\|=1} \|Tv\|.$$

We say that T is *bounded* if $\|T\| < \infty$. Note that for every vector v we have $\|Tv\| \leq \|T\|\,\|v\|$, as can be seen by replacing v with $\frac{1}{\|v\|}v$ if $v \neq 0$.

Let $\mathcal{B}(H)$ be the set of all bounded linear operators. It is straightforward to see that the operator norm actually satisfies the axioms of a norm, so $\mathcal{B}(H)$ is a normed vector space (see Exercise 12.2).

Lemma 12.3.1 *The space $\mathcal{B}(H)$ is complete; i.e., it is a Banach space.*

Proof: Let (T_n) be a Cauchy sequence in $\mathcal{B}(H)$. We show first that for every $w \in H$ the sequence $T_n w$ converges. So fix $w \in H$. Replacing w by a scalar multiple if necessary, we can assume that $\|w\| = 1$. For $m, n \in \mathbb{N}$ we have

$$\|T_m w - T_n w\| \leq \sup_{\|v\|=1} \|T_m v - T_n v\| = \|T_m - T_n\|.$$

It follows that the sequence $(T_n w)$ is a Cauchy sequence in H, hence convergent. We define

$$Tw \overset{\text{def}}{=} \lim_n T_n w.$$

This defines a linear operator T on H.

Let $\varepsilon > 0$ and choose $n_0 \in \mathbb{N}$ such that for all $m, n \geq n_0$ we have $\|T_m - T_n\| < \varepsilon$. For $n \geq n_0$ and every $w \in H$ with $\|w\| = 1$ we have

$$\|Tw - T_n w\| = \lim_m \|T_m w - T_n w\| \leq \varepsilon.$$

Therefore, $\|Tw\| \leq \|Tw - T_n w\| + \|T_n w\| \leq \varepsilon + \|T_n\|$, which implies that T is bounded, so $T \in \mathcal{B}(H)$. Furthermore, the above estimate being uniform in w implies that the sequence (T_n) actually converges to T in $\mathcal{B}(H)$. $\qquad\square$

Lemma 12.3.2 *A linear operator T on a Hilbert space is continuous if and only if it is bounded.*

Proof: A linear operator is continuous if and only if it is continuous at zero, i.e., iff Tv_n tends to zero whenever v_n tends to zero. Suppose T is continuous and assume that there is a sequence $(v_n)_{n \in \mathbb{N}}$ with $\|Tv_n\| \to \infty$. Suppose $Tv_n \neq 0$ for every n. Set

$$w_n \overset{\text{def}}{=} \frac{1}{\|Tv_n\|} v_n.$$

Then w_n tends to zero and hence Tw_n tends to zero. So

$$1 = \frac{\|Tv_n\|}{\|Tv_n\|} = \|Tw_n\|$$

tends to zero, a contradiction! It follows that such a sequence v_n does not exist, and so $\|T\| < \infty$.

Conversely, assume T is bounded. Then if $v_n \to 0$, we get $\|Tv_n\| \leq \|T\| \|v_n\|$, and this also tends to zero; so T is continuous. $\qquad\square$

Lemma 12.3.3 *Let $\alpha : H \to \mathbb{C}$ be linear and continuous. Then there exists a unique $w_0 \in H$ such that*

$$\alpha(v) = \langle v, w_0 \rangle$$

for every $v \in H$.

Proof: We show uniqueness first. Suppose there is a second w_0'. Then $\langle v, w_0 - w_0' \rangle = 0$ for every $v \in H$, so this holds in particular for $v = w_0 - w_0'$, which implies $w_0 - w_0' = 0$ as claimed.

Suppose $\alpha \neq 0$, because otherwise, the assertion will be trivial. Let $V = \ker(\alpha) = \{v \in H \mid \alpha(v) = 0\}$ be the *kernel* of α. Then V is a closed subspace of H. Let $U = V^\perp = \{w \in H \mid \langle w, v \rangle = 0 \ \forall v \in V\}$, the *orthogonal space*. Then α maps U isomorphically to $\mathbb{C}$, so there is a unique $w \in U$ with $\alpha(w) = 1$. Let $v \in H$ be arbitrary. Let $\lambda = \alpha(v) \in \mathbb{C}$. Then $v - \lambda w \in \ker(\alpha)$, and hence $\langle v - \lambda w, w \rangle = 0$ or

$$\langle v, w \rangle = \lambda \langle w, w \rangle.$$

Set $w_0 \overset{\text{def}}{=} \frac{1}{\langle w, w \rangle} w$. Then $\langle v, w_0 \rangle = \lambda = \alpha(v)$. $\qquad\square$

Next let $T \in \mathcal{B}(H)$ and fix $w \in H$. Then the linear map $\alpha \colon H \to \mathbb{C}$ given by $\alpha(v) = \langle Tv, w \rangle$ is continuous; hence there exists a unique $T^*w \in H$ such that $\langle Tv, w \rangle = \langle v, T^*w \rangle$. Since this holds for every $w \in H$, we get a map $w \mapsto T^*w$, and it is easy to see that this map T^* is linear. It is called the *adjoint* of T. We will show that it is bounded. For this let $v, w \in H$. Then

$$| \langle v, T^*w \rangle | \; = \; | \langle Tv, w \rangle | \; \le \; \|Tv\| \, \|w\| \; \le \; C \, \|v\| \, \|w\|$$

for $C = \|T\|$. In particular, if we choose $v = T^*w$, we get

$$\|T^*w\|^2 \; \le \; C \, \|T^*w\| \, \|w\|,$$

which implies $\|T^*w\| \le C \, \|w\|$.

Proposition 12.3.4 *The map $T \mapsto T^*$ defines a norm-preserving involution on $\mathcal{B}(H)$; i.e., for $S, T \in \mathcal{B}(H)$ and $\lambda \in \mathbb{C}$ we have*

(a) $(T^*)^* = T$,

(b) $(S + T)^* = S^* + T^*$,

(c) $(ST)^* = T^*S^*$,

(d) $(\lambda T)^* = \bar{\lambda} T^*$,

(e) $\|T\| = \|T^*\|$.

Proof: The points (a), (b), and (d) are trivial. For (c) compute

$$\langle v, (ST)^* w \rangle \; = \; \langle STv, w \rangle \; = \; \langle Tv, S^*w \rangle \; = \; \langle v, T^*S^*w \rangle .$$

To see (e), note that we have shown $\|T^*\| \le \|T\|$ above. Replacing T with T^* implies the claim. $\square$

Let $T \in \mathcal{B}(H)$ and let (e_j) be an orthonormal basis of H. The *Hilbert-Schmidt norm* $\|T\|_{HS}$ of T is defined by

$$\|T\|_{HS}^2 \; \overset{\text{def}}{=} \; \sum_j \langle Te_j, Te_j \rangle .$$

This number is ≥ 0 but can be $+\infty$. We have to show that it does not depend on the choice of the orthonormal basis. For this recall that Theorem 2.3.2 implies that for every orthonormal basis $(\phi_\alpha)_\alpha$ and all $v, w \in H$ we have

$$\langle v, w \rangle = \sum_\alpha \langle v, \phi_\alpha \rangle \langle \phi_\alpha, w \rangle.$$

So let ϕ_α be another orthonormal basis. Not knowing the independence yet, we denote the Hilbert-Schmidt norm attached to the orthonormal basis (e_j) by $\|T\|_{HS,(e)}$. Then

$$
\begin{aligned}
\|T\|^2_{HS,(e)} &= \sum_j \langle Te_j, Te_j \rangle \\
&= \sum_j \sum_\alpha \langle Te_j, \phi_\alpha \rangle \langle \phi_\alpha, Te_j \rangle \\
&= \sum_\alpha \sum_j \langle Te_j, \phi_\alpha \rangle \langle \phi_\alpha, Te_j \rangle \\
&= \sum_\alpha \sum_j \langle e_j, T^*\phi_\alpha \rangle \langle T^*\phi_\alpha, e_j \rangle \\
&= \sum_\alpha \langle T^*\phi_\alpha, T^*\phi_\alpha \rangle = \|T^*\|_{HS,(\phi)}.
\end{aligned}
$$

The interchange of order of summation is justified, since all summands are positive. For $(\phi) = (e)$ this in particular implies $\|T^*\|_{HS,(\phi)} = \|T\|_{HS,(\phi)}$, so that

$$\|T\|_{HS,(e)} = \|T^*\|_{HS,(\phi)} = \|T\|_{HS,(\phi)}.$$

This gives the desired independence, so $\|T\|_{HS}$ is well-defined. We say that the operator T is a *Hilbert-Schmidt operator*, if $\|T\|_{HS} < \infty$.

Lemma 12.3.5 *For every bounded operator T on H,*

$$\|T\| \le \|T\|_{HS}.$$

For every unitary operator U we have $\|UT\|_{HS} = \|TU\|_{HS} = \|T\|_{HS}$.

Proof: Let $v \in H$ with $\|v\| = 1$. Then there is an orthonormal basis (e_j) with $e_1 = v$. We get

$$\|Tv\|^2 = \|Te_1\|^2 \le \sum_j \|Te_j\|^2 = \|T\|^2_{HS}.$$

The invariance under multiplication by unitary operators is clear, since (Ue_j) is an orthonormal basis when (e_j) is. $\square$

The main example we are interested in is the following. Recall $L^2(\mathbb{R})$, which is the Hilbert space completion of $L^2_{bc}(\mathbb{R})$ as well as that of $C_c(\mathbb{R})$. Let k be a continuous bounded function on $\mathbb{R}^2$ and suppose

$$\int_{\mathbb{R}} \int_{\mathbb{R}} |k(x,y)|^2 \, dx \, dy \; < \; \infty.$$

This double integral is to be understood as follows. We assume that for every $x \in \mathbb{R}$ the integral $\int_{\mathbb{R}} |k(x,y)|^2 \, dy$ exists and defines a continuous function on $\mathbb{R}$ that is integrable, and the same with x and y interchanged. Under these circumstances we call k an L^2-kernel.

Proposition 12.3.6 *Suppose $k(x,y)$ is an L^2-kernel on $\mathbb{R}$. For $\varphi \in C_c(\mathbb{R})$ define*

$$K\varphi(x) \; \overset{\text{def}}{=} \; \int_{\mathbb{R}} k(x,y)\varphi(y) \, dy.$$

Then $K\varphi$ lies in $L^2_{bc}(\mathbb{R})$, and K extends to a Hilbert-Schmidt operator $K : L^2(\mathbb{R}) \to L^2(\mathbb{R})$ with

$$\|K\|^2_{HS} \; = \; \int_{\mathbb{R}} \int_{\mathbb{R}} |k(x,y)|^2 \, dx \, dy.$$

Proof: The function $K\varphi$ is clearly continuous and bounded. We use the Cauchy-Schwartz inequality to estimate

$$\begin{aligned}
\|K\varphi\|^2 \; &= \; \int_{\mathbb{R}} |K\varphi(x)|^2 \, dx \\
&= \; \int_{\mathbb{R}} \left| \int_{\mathbb{R}} k(x,y)\varphi(y) \, dy \right|^2 dx \\
&\leq \; \int_{\mathbb{R}} \int_{\mathbb{R}} |k(x,y)|^2 \, dx \, dy \int_{\mathbb{R}} |\varphi(y)|^2 \, dy \\
&= \; \int_{\mathbb{R}} \int_{\mathbb{R}} |k(x,y)|^2 \, dx \, dy \, \|\varphi\|^2 .
\end{aligned}$$

So K extends to a bounded operator on $L^2(\mathbb{R})$. Let (e_j) be an orthonormal basis of $L^2(\mathbb{R})$. Then

$$
\begin{aligned}
\|K\|_{HS} &= \sum_j \langle Ke_j, Ke_j \rangle \\
&= \sum_j \int_{\mathbb{R}} Ke_j(x)\overline{Ke_j(x)}\, dx \\
&= \sum_j \int_{\mathbb{R}} \int_{\mathbb{R}} k(x,y)e_j(y)\, dy \overline{\int_{\mathbb{R}} k(x,y)e_j(y)\, dy}\, dx \\
&= \sum_j \int_{\mathbb{R}} \langle k(x,.), e_j \rangle \, \langle e_j, k(x,.) \rangle \, dx \\
&= \int_{\mathbb{R}} \sum_j \langle k(x,.), e_j \rangle \, \langle e_j, k(x,.) \rangle \, dx \\
&= \int_{\mathbb{R}} \langle k(x,.), k(x,.) \rangle \, dx \\
&= \int_{\mathbb{R}} \int_{\mathbb{R}} |k(x,y)|^2 \, dx\, dy.
\end{aligned}
$$

$\square$

12.4   The Plancherel Theorem for $\mathcal{H}$

Let G be a locally compact group and let $f \in C_c(G)$. Fix a Haar measure on G. For a unitary representation (π, V_π) of G we define, formally at first,

$$
\pi(f) = \int_G f(x)\pi(x)\, dx
$$

as the unique linear operator on V_π that satisfies

$$
\langle \pi(f)v, w \rangle = \int_G f(x)\, \langle \pi(x)v, w \rangle \, dx
$$

for all $v, w \in V_\pi$. We claim that $\pi(f)$ is bounded and that

$$
\|\pi(f)\| \le \|f\|_1 = \int_G |f(x)|\, dx.
$$

To prove this we employ the Cauchy-Schwarz inequality to get

$$
\begin{aligned}
|\langle \pi(f)v, w\rangle| &\leq \int_G |f(x)|\,|\langle \pi(x)v, w\rangle|\,dx \\
&\leq \int_G |f(x)|\,\|v\|\,\|w\|\,dx \\
&= \|f\|_1\,\|v\|\,\|w\|.
\end{aligned}
$$

For $w = \pi(f)v$ this gives $\|\pi(f)v\|^2 \leq \|f\|_1\,\|v\|\,\|\pi(f)v\|$, which implies
the claim. $\square$

We have an identification $\mathcal{H} \cong \mathbb{R}^3$. Now let $\mathcal{S}(\mathbb{R}^3)$ be the space of
Schwartz functions on $\mathbb{R}^3$. This is the space of all infinitely differentiable functions f on $\mathbb{R}^3$ such that for all $k, m, n \in \mathbb{N}_0$ and every
polynomial $P(x, y, z)$ the function

$$
P(x,y,z)\left(\frac{\partial}{\partial x}\right)^k \left(\frac{\partial}{\partial y}\right)^m \left(\frac{\partial}{\partial z}\right)^n f(x,y,z)
$$

on $\mathbb{R}^3$, is bounded.

Using the above identification, we interpret $f \in \mathcal{S}(\mathbb{R}^3)$ as a function
on $\mathcal{H}$, and we write $\mathcal{S}(\mathcal{H})$ for this space of functions.

Theorem 12.4.1 *(Plancherel theorem)*
*Let $f \in \mathcal{S}(\mathcal{H})$, For every $t \in \mathbb{R}^\times$ the operator $\pi_t(f)$ is a Hilbert-
Schmidt operator, and we have*

$$
\int_{\mathbb{R}^\times} \|\pi_t(f)\|_{HS}^2\, |t|\,dt = \int_{\mathcal{H}} |f(h)|^2\,dh.
$$

Note that the one-dimensional representations do not occur in the
Plancherel theorem. We say that they have *Plancherel measure zero.*

Proof: Let $t \in \mathbb{R}^\times$, $f \in \mathcal{S}(\mathcal{H})$, and $\varphi \in L^2(\mathbb{R})$. Then

$$
\begin{aligned}
\pi_t(f)\varphi(x) &= \int_{\mathbb{R}^3} f(a,b,c)\pi_1(a, bt, ct)\varphi(x)\,da\,db\,dc \\
&= \int_{\mathbb{R}^3} f(a,b,c)e^{2\pi i t(bx+c)}\varphi(x+a)\,da\,db\,dc \\
&= \int_{\mathbb{R}^3} f(a-x,b,c)e^{2\pi i t(bx+c)}\varphi(a)\,da\,db\,dc \\
&= \int_{\mathbb{R}} k(x,y)\varphi(y)\,dy,
\end{aligned}
$$

where

$$
\begin{aligned}
k(x,y) &= \int_{\mathbb{R}^2} f(y-x,b,c)\, e^{2\pi i t(bx+c)}\, db\, dc \\
&= \int_{\mathbb{R}} \mathcal{F}_2 f(y-x,-tx,c)\, e^{2\pi i t c}\, dc \\
&= \mathcal{F}_3 \mathcal{F}_2 f(y-x,-tx,-t),
\end{aligned}
$$

where $\mathcal{F}_2$ and $\mathcal{F}_3$ denote the Fourier transforms with respect to the second and third variable, respectively. The kernel k is bounded and continuous. The function $g = \mathcal{F}_3 \mathcal{F}_2 f$ lies in $\mathcal{S}(\mathcal{H})$ again and we have

$$
\int_{\mathbb{R}} |k(x,y)|^2 dy = \int_{\mathbb{R}} |g(y,-tx,-t)|^2 dy
$$

as well as

$$
\int_{\mathbb{R}} k(x,y)|^2 dx = \int_{\mathbb{R}} |g(y-x,-tx,-t)|^2 dx,
$$

which implies that k is an L^2-kernel. By Proposition 12.3.6 the operator $\pi_t(f)$ is Hilbert-Schmidt. The same proposition together with the Plancherel theorem for the Fourier transform gives

$$
\begin{aligned}
\int_{\mathbb{R}^\times} \|\pi_t(f)\|_{HS}^1 |t|\, dt &= \int_{\mathbb{R}^\times} \int_{\mathbb{R}^2} |g(y,-tx,-t)|^2 dx\, dy\, |t|\, dt \\
&= \int_{\mathbb{R}^\times} \int_{\mathbb{R}^2} |g(y,x,t)|^2 dx\, dy\, dt \\
&= \int_{\mathbb{R}^\times} \int_{\mathbb{R}^2} |f(y,x,t)|^2 dx\, dy\, dt.
\end{aligned}
$$

$\square$

12.5 A Reformulation

The Plancherel Theorem for $\mathcal{H}$ provides a decomposition of the unitary $\mathcal{H} \times \mathcal{H}$ representation on $L^2(\mathcal{H})$ given by

$$
R(h_1, h_2)\varphi(h) \overset{\mathrm{def}}{=} \varphi(h_1^{-1} h h_2).
$$

This representation will be decomposed not as a direct sum but as a direct integral. The general concept of direct integrals [3] requires

Lebesgue integration and is therefore beyond the scope of the book. In the particular case of the Heisenberg group, however, we can give a simplified construction that does the job.

For a given Hilbert space V we consider the space $\mathrm{HS}(V)$ of all Hilbert-Schmidt operators on V. We choose an orthonormal basis (e_j), and for $S, T \in \mathrm{HS}(V)$ we define

$$\langle S, T \rangle \overset{\text{def}}{=} \sum_j \langle Se_j, Te_j \rangle.$$

Lemma 12.5.1 *For $S, T \in \mathrm{HS}(V)$ the sum defining $\langle S, T \rangle$ converges, and its value does not depend on the choice of the orthonormal basis. This defines an inner product on $\mathrm{HS}(V)$. The space $\mathrm{HS}(V)$ is complete with respect to this inner product, i.e., $\mathrm{HS}(V)$ is a Hilbert space.*

Proof: The Cauchy-Schwarz inequality implies $\sum_j |\langle Se_j, Te_j \rangle| \leq \sum_j \|Se_j\| \|Te_j\|$, and since the sequences $(\|Se_j\|)_{j \in \mathbb{N}}$ and $(\|Te_j\|)_{j \in \mathbb{N}}$ are in $\ell^2(\mathbb{N})$, the latter sum converges. The independence of the choice of the orthonormal basis is shown similar to the independence of the norm. The axioms for an inner product are easily established. It remains to show that $\mathrm{HS}(V)$ is complete. To this end let (S_n) be a Cauchy sequence in $\mathrm{HS}(V)$. By Lemma 12.3.5 it follows that (S_n) is a Cauchy sequence in $\mathcal{B}(V)$ as well and thus has a limit $S \in \mathcal{B}(V)$.

Let $\varepsilon > 0$ and let $n_0 \in \mathbb{N}$ be such that for all $m, n \geq n_0$ we have $\|S_n - S_m\|_{HS}^2 < \varepsilon$. Let (e_j) be an orthonormal basis of V. Then for every $n \geq n_0$,

$$\|S_n - S\|_{HS}^2 = \sum_j \|S_n e_j - S e_j\|^2 = \sum_j \lim_m \|S_n e_j - S_m e_j\|^2.$$

For every $j_0 \in \mathbb{N}$ we have

$$\sum_{j \leq j_0} \lim_m \|S_n e_j - S_m e_j\|^2 = \lim_m \sum_{j \leq j_0} \|S_n e_j - S_m e_j\|^2$$

$$\leq \limsup_m \sum_j \|S_n e_j - S_m e_j\|^2$$

$$= \limsup_m \|S_n - S_m\|_{HS}^2 \leq \varepsilon.$$

By letting j_0 tend to infinity we get $\|S_n - S\|_{HS}^2 \leq \varepsilon$. Varying ε implies that the sequence S_n tends to S in $\mathrm{HS}(V)$. $\qquad\square$

Let (π, V_π) be a unitary representation of a locally compact group G. On the Hilbert space $\mathrm{HS}(V_\pi)$ we can define a representation π_{HS} of $G \times G$ by

$$\pi_{\mathrm{HS}}(g_1, g_2)T \overset{\mathrm{def}}{=} \pi(g_1)T\pi(g_2^{-1}).$$

Since $\pi(g_1)$ and $\pi(g_2^{-1})$ are unitary operators on V_π, it follows that π_{HS} is a unitary representation of $G \times G$ on $\mathrm{HS}(V_\pi)$. In particular, for each $t \in \mathbb{R}^\times$ we get a representation $\pi_{t,\mathrm{HS}}$ of $\mathcal{H} \times \mathcal{H}$ on the space $\mathrm{HS}(L^2(\mathbb{R}))$.

Let H be a Hilbert space. We are going to define the space

$$L^2(\mathbb{R}^\times, H, |t|dt).$$

We first define the space $C_c(\mathbb{R}^\times, H)$ as the space of all continuous functions $\varphi \colon \mathbb{R}^\times \to H$ with compact support. On this space we introduce an inner product by

$$\langle \varphi, \psi \rangle \overset{\mathrm{def}}{=} \int_{\mathbb{R}^\times} \langle \varphi(t), \psi(t) \rangle \, |t|dt,$$

where the inner product on the right-hand side is that of the Hilbert space H. Finally, $L^2(\mathbb{R}^\times, H, |t|dt)$ is the Hilbert completion of $C_c(\mathbb{R}^\times, H)$. This is an example of a direct integral of Hilbert spaces, the idea being that for each $t \in \mathbb{R}^\times$ one copy H_t of H is taken, and these are integrated over $\mathbb{R}^\times$ to form a new Hilbert space.

We now consider this construction in the special case

$$H = \mathrm{HS}(L^2(\mathbb{R})).$$

Lemma 12.5.2 *On the space $L^2(\mathbb{R}^\times, \mathrm{HS}(L^2(\mathbb{R})), |t|dt)$ we have a unitary representation Π of $\mathcal{H} \times \mathcal{H}$ given by*

$$\Pi(h_1, h_2)\varphi(t) \overset{\mathrm{def}}{=} \pi_{t,\mathrm{HS}}(h_1, h_2)\varphi(t) = \pi_t(h_1)\varphi(t)\pi_t(h_2^{-1}).$$

Proof: To show that Π is unitary, we compute

$$\begin{aligned}
\|\Pi(h_1, h_2)\varphi\|^2 &= \int_{\mathbb{R}^\times} \|\Pi(h_1, h_2)\varphi(t)\|^2_{\mathrm{HS}} \, |t|dt \\
&= \int_{\mathbb{R}^\times} \left\|\pi_t(h_1)\varphi(t)\pi_t(h_2^{-1})\right\|^2_{\mathrm{HS}} \, |t|dt \\
&= \int_{\mathbb{R}^\times} \|\varphi(t)\|^2_{\mathrm{HS}} \, |t|dt = \|\varphi\|^2.
\end{aligned}$$

□

This representation is called the *direct integral* of the representations $\pi_{t,\mathrm{HS}}$ and is written

$$\Pi = \int_{\mathbb{R}^\times} \pi_{t,\mathrm{HS}} |t| dt.$$

Theorem 12.5.3 *The map*

$$\Psi : \mathcal{S}(\mathcal{H}) \;\to\; L^2(\mathbb{R}^\times, \mathrm{HS}(L^2(\mathbb{R}), |t| dt),$$
$$f \;\mapsto\; \Psi(f),$$

with $\Psi(f)(t) \overset{\text{def}}{=} \pi_t(f)$ *is* $\mathcal{H} \times \mathcal{H}$*-equivariant, injective, and has a dense image. It satisfies* $\|\Psi(f)\| = \|f\|$*. It extends to an isomorphism of Hilbert spaces*

$$\Psi : L^2(\mathcal{H}) \;\to\; L^2(\mathbb{R}^\times, \mathrm{HS}(L^2(\mathbb{R}), |t| dt),$$

which satisfies $\Psi(R(h_1, h_2)f) = \Pi(h_1, h_2)\Psi(f)$*. In other words, the regular representation* R *of* $\mathcal{H}$ *decomposes,*

$$R \cong \int_{\mathbb{R}^\times} \pi_{t,\mathrm{HS}} |t| dt.$$

Proof: To see that Ψ is $\mathcal{H} \times \mathcal{H}$ equivariant we compute

$$
\begin{aligned}
\Psi(R(h_1, h_2)f)(t) &= \pi_t(R(h_1, h_2)f) \\
&= \int_G R(h_1, h_2)f(x)\,\pi_t(x)\,dx \\
&= \int_G f(h_1^{-1}xh_2)\,\pi_t(x)\,dx \\
&= \int_G f(x)\,\pi_t(h_1 x h_2^{-1})\,dx \\
&= \pi_t(h_1)\int_G f(x)\,\pi_t(x)\,dx\,\pi_t(h_2^{-1}) \\
&= \pi_t(h_1)\pi_t(f)\pi_t(h_2^{-1}) \\
&= \Pi(h_1, h_2)\Psi(f).
\end{aligned}
$$

Further, since $\pi_t(f)$ has the kernel

$$k_t(x, y) = \mathcal{F}_3\mathcal{F}_2 f(y - x, -tx, -t),$$

the assumption $\Psi(f) = 0$ implies $k_t(x, y) = 0$ for all x, y, t, and this implies $f = 0$. So Ψ is injective. The fact that Ψ preserves norms is the Plancherel theorem. Finally, to see that Ψ has dense image, note that the image contains all kernels $k_t(x, y)$ that are smooth and compactly supported in $\mathbb{R}^2 \times \mathbb{R}^\times$. This implies the density. $\qquad \square$

12.6 Exercises

Exercise 12.1 Show that for an LCA group A the unitary dual can be identified with the dual group.

(Use Lemma 12.2.1 to see that every irreducible unitary representation of A is one-dimensional. Identify its isomorphism class with its "central" character.)

Exercise 12.2 Show that the operator norm satisfies the axioms of a norm as in Lemma 2.1.1.

Exercise 12.3 Let H be a Hilbert space. An operator T on H is called a *finite rank operator* if the image $T(H)$ is finite-dimensional. Show that the operators of finite rank form a dense subspace of $\mathrm{HS}(H)$.

Exercise 12.4 Let G be a locally compact group. For $f, g \in C_c(G)$ define their *convolution* by

$$f * g(x) \stackrel{\mathrm{def}}{=} \int_G f(y)g(y^{-1}x) \, dy.$$

Let π be a unitary representation of G. Show that

$$\pi(f * g) = \pi(f)\pi(g).$$

Exercise 12.5 Show that the set of (a, b, c), where a, b, c are integers, forms a closed subgroup Γ of $\mathcal{H}$. Show that the quotient $\mathcal{H}/\Gamma$ is compact.

Appendix A

The Riemann Zeta Function

We now give the analytic continuation and the functional equation of the Riemann zeta function, which is based on the functional equation of the theta series. First we need the gamma function:

For $\mathrm{Re}(s) > 0$ the integral

$$\Gamma(s) = \int_0^\infty t^{s-1} e^{-t} dt$$

converges and gives a holomorphic function in that range. We integrate by parts to get for $\mathrm{Re}(s) > 0$,

$$\Gamma(s+1) = \int_0^\infty t^s e^{-t} dt = \int_0^\infty s t^{s-1} e^{-t} dt = s\Gamma(s),$$

i.e.,

$$\Gamma(s) = \frac{\Gamma(s+1)}{s}.$$

In the last equation the right-hand-side gives a meromorphic function on $\mathrm{Re}(s) > -1$, and thus $\Gamma(s)$ extends meromorphically to that range. But again the very same equation extends $\Gamma(s)$ to $\mathrm{Re}(s) > -2$, and so on. We find that $\Gamma(s)$ extends to a meromorphic function on the entire plane that is holomorphic except for simple poles at $s = 0, -1, -2, \ldots$.

Recall from Section 3.6 the theta series

$$\Theta(t) = \sum_{k \in \mathbb{Z}} e^{-t\pi k^2}, \text{ for } t > 0,$$

175

which satisfies

$$\Theta(t) = \frac{1}{\sqrt{t}} \Theta\left(\frac{1}{t}\right),$$

as was shown in Theorem 3.7.1. We now introduce the Riemann zeta function:

Lemma A.1 *For* $\mathrm{Re}(s) > 1$ *the series*

$$\zeta(s) = \sum_{n=1}^{\infty} \frac{1}{n^s}$$

converges absolutely and defines a holomorphic function there. This function is called the Riemann zeta function.

Proof: Since the summands $1/n^s$ are entire functions, it needs to be shown only that the series $\sum_{n=1}^{\infty} |n^{-s}|$ converges locally uniformly in $\mathrm{Re}(s) > 1$. In that range we compute

$$\frac{1}{\mathrm{Re}(s) - 1} = \left.\frac{x^{-\mathrm{Re}(s)+1}}{1 - \mathrm{Re}(s)}\right|_1^{\infty} = \int_1^{\infty} x^{-\mathrm{Re}(s)} dx$$

$$= \int_2^{\infty} (x-1)^{-\mathrm{Re}(s)} dx \geq \int_2^{\infty} [x]^{-\mathrm{Re}(s)} dx$$

$$= \sum_{n=2}^{\infty} n^{-\mathrm{Re}(s)} = \sum_{n=2}^{\infty} \left|\frac{1}{n^s}\right|,$$

where for $x \in R$ the number $[x]$ is the largest integer k that satisfies $k \leq x$. The lemma follows. $\qquad\square$

Theorem A.2 *(The functional equation of the Riemann zeta function)*

The Riemann zeta function $\zeta(s)$ *extends to a meromorphic function on* $\mathbb{C}$, *holomorphic up to a simple pole at* $s = 1$, *and the function*

$$\xi(s) = \pi^{-s/2} \Gamma\left(\frac{s}{2}\right) \zeta(s)$$

satisfies

$$\xi(s) = \xi(1-s)$$

for every $s \in \mathbb{C}$.

Proof: Note that the expression dt/t is invariant under the substitution $t \mapsto ct$ for $c > 0$ and up to sign under $t \mapsto 1/t$. Using these facts, we compute for $\mathrm{Re}(s) > 1$,

$$\xi(s) = \zeta(s)\Gamma\left(\frac{s}{2}\right)\pi^{-s/2} = \sum_{n=1}^{\infty}\int_0^{\infty} n^{-s}t^{s/2}\pi^{-s/2}e^{-t}\frac{dt}{t}$$

$$= \sum_{n=1}^{\infty}\int_0^{\infty}\left(\frac{t}{n^2\pi}\right)^{s/2}e^{-t}\frac{dt}{t} = \sum_{n=1}^{\infty}\int_0^{\infty}t^{s/2}e^{-n^2\pi t}\frac{dt}{t}$$

$$= \int_0^{\infty}t^{s/2}\frac{1}{2}(\Theta(t)-1)\frac{dt}{t}.$$

We split this integral into a sum of an integral over $(0,1)$ and an integral over $(1,\infty)$. The latter one,

$$\int_1^{\infty}t^{s/2}\frac{1}{2}(\Theta(t)-1)\frac{dt}{t},$$

is an entire function, since the function $t \mapsto \Theta(t) - 1$ is rapidly decreasing at ∞. The other summand is

$$\int_0^1 t^{s/2}\frac{1}{2}(\Theta(t)-1)\frac{dt}{t} = \int_1^{\infty}t^{-s/2}\frac{1}{2}\left(\Theta\left(\frac{1}{t}\right)-1\right)\frac{dt}{t}$$

$$= \int_1^{\infty}t^{-s/2}\frac{1}{2}\left(\sqrt{t}\Theta(t)-1\right)\frac{dt}{t}$$

$$= \int_1^{\infty}t^{-s/2}\frac{1}{2}\left(\sqrt{t}(\Theta(t)-1)+\sqrt{t}-1\right)\frac{dt}{t},$$

which equals the sum of the entire function

$$\int_1^{\infty}t^{(1-s)/2}\frac{1}{2}(\Theta(t)-1)\frac{dt}{t}$$

and

$$\frac{1}{2}\int_1^{\infty}t^{(1-s)/2}\frac{dt}{t} - \frac{1}{2}\int_1^{\infty}t^{-s/2}\frac{dt}{t} = \frac{1}{s-1}-\frac{1}{s}.$$

Summarizing, we get

$$\xi(s) = \int_1^{\infty}\left(t^{\frac{s}{2}}+t^{\frac{1-s}{2}}\right)\frac{1}{2}(\Theta(t)-1)\frac{dt}{t}-\frac{1}{s}-\frac{1}{1-s}.$$

$\square$

Using the functional equation and knowing the locations of the poles of the Γ-function, we can see that the Riemann zeta function has zeros at the even negative integers $-2, -4, -6, \ldots$, called the *trivial zeros*. It can be shown that all other zeros are in the strip $0 < \mathrm{Re}(s) < 1$. The up to now unproven *Riemann hypothesis* states that all nontrivial zeros should be in the set $\mathrm{Re}(s) = \frac{1}{2}$. This would have deep consequences about the distribution of primes through the prime number theorem [13].

This technique for constructing the analytic continuation of the zeta function dates back to Riemann, and can be applied to other Dirichlet series as well.

Appendix B

Haar Integration

Let G be an LC group. We here give the proof of the existence of a Haar integral.

Theorem B.1 *There exists a non-zero invariant integral I of G. If I' is a second non-zero invariant integral, then there is a number $c > 0$ such that $I' = cI$.*

For the uniqueness part of the theorem we say that the invariant integral is *unique up to scaling*.

The idea of the proof resembles the construction of the Riemann integral on $\mathbb{R}$. To construct the Riemann integral of a positive function one finds a step function that dominates the given function and adds the lengths of the intervals needed multiplied by the values of the dominating function. Instead of characteristic functions of intervals one could also use translates of a given continuous function with compact support, and this is exactly what is done in the general situation.

Proof of the Theorem: For the existence part, let $C_c^+(G)$ be the set of all $f \in C_c(G)$ with $f \geq 0$. For $f, g \in C_c(G)$ with $g \neq 0$ there are $c_j > 0$ and $s_j \in G$ such that

$$f(x) \leq \sum_{j=1}^{n} c_j g(s_j^{-1} x).$$

179

Let $(f : g)$ denote

$$\inf \left\{ \sum_{j=1}^{n} c_j \;\middle|\; \begin{array}{l} c_1, \ldots, c_n > 0 \text{ and there are } s_1, \ldots, s_n \in G \\ \text{such that } f(x) \le \sum_{j=1}^{n} c_j g(s_j x) \end{array} \right\}.$$

Lemma B.2 *For $f, g, h \in C_c^+(G)$ with $g \ne 0$ we have*

(a) $(L_s f : g) = (f : g)$ *for every $s \in G$,*

(b) $(f + h : g) \le (f : g) + (h : g)$,

(c) $(\lambda f : g) = \lambda (f, g)$ *for $\lambda \ge 0$,*

(d) $f \le h \;\Rightarrow\; (f : g) \le (h : g)$,

(e) $(f : h) \le (f : g)(g : h)$ *if $h \ne 0$, and*

(f) $(f : g) \ge \frac{\max f}{\max g}$, *where $\max f = \max\{f(x)|x \in G\}$.*

Proof: The items (a) to (d) are trivial. For item (e) let $f(x) \le \sum_j c_j g(s_j x)$ and $g(y) \le \sum_k d_k h(t_k y)$; then

$$f(x) \le \sum_{j,k} c_j d_k h(t_k s_j x),$$

so that $(f : h) \le \sum_j c_j \sum_k d_k$.

For (f) choose $x \in G$ with $\max f = f(x)$; then

$$\max f = f(x) \le \sum_j c_j g(s_j x) \le \sum_j c_j \max g.$$

$\square$

Fix some $f_0 \in C_c^+(G)$, $f_0 \ne 0$. For $f, \varphi \in C_c^+(G)$ with $\varphi \ne 0$ let

$$J(f, \varphi) = J_{f_0}(f, \varphi) = \frac{(f : \varphi)}{(f_0 : \varphi)}.$$

Lemma B.3 *For $f, h, \varphi \in C_c^+(G)$ with $f, \varphi \ne 0$ we have*

(a) $\frac{1}{(f_0 : f)} \le J(f, \varphi) \le (f : f_0)$,

(b) $J(L_s f, \varphi) = J(f, \varphi)$ *for every* $s \in G$,

(c) $J(f + h, \varphi) \leq J(f, \varphi) + J(h, \varphi)$, *and*

(d) $J(\lambda f, \varphi) = \lambda J(f, \varphi)$ *for every* $\lambda \geq 0$.

Proof: This follows from the last lemma. □

The function $f \mapsto J(f, \varphi)$ does not give an integral, since it is not additive but only subadditive. However, as the support of φ shrinks it will become asymptotically additive, as the next lemma shows.

Lemma B.4 *Given* $f_1, f_2 \in C_c^+(G)$ *and* $\varepsilon > 0$ *there is a neighborhood* V *of the unit in* G *such that*

$$J(f_1, \varphi) + J(f_2, \varphi) \leq J(f_1 + f_2, \varphi)(1 + \varepsilon)$$

holds for every $\varphi \in C_c^+(G)$, $\varphi \neq 0$ *with support contained in* V.

Proof: Choose $f' \in C_c^+(G)$ such that f' is identically equal to 1 on the support of $f_1 + f_2$. For the existence of such a function see Exercise 8.2. Let $\delta, \varepsilon > 0$ be arbitrary and set

$$f = f_1 + f_2 + \delta f', \quad h_1 = \frac{f_1}{f}, \quad h_2 = \frac{f_2}{f},$$

where it is understood that $h_j = 0$ where $f = 0$. It follows that $h_j \in C_c^+(G)$.

Choose a neighborhood V of the unit such that $|h_j(x) - h_j(y)| < \varepsilon/2$ whenever $x^{-1}y \in V$. If $\text{supp}(\varphi) \subset V$ and $f(x) \leq \sum_k c_k \varphi(s_k x)$, then $\varphi(s_k x) \neq 0$ implies

$$|h_j(x) - h_j(s_k^{-1})| < \frac{\varepsilon}{2},$$

and

$$f_j(x) = f(x) h_j(x) \leq \sum_k c_k \varphi(s_k x) h_j(x)$$

$$\leq \sum_k c_k \varphi(s_k x) \left(h_j(s_k^{-1}) + \frac{\varepsilon}{2} \right),$$

so that

$$(f_j : \varphi) \leq \sum_k c_k \left(h_j(s_k^{-1}) + \frac{\varepsilon}{2} \right),$$

and so

$$(f_1 : \varphi) + (f_2 : \varphi) \leq \sum_k c_k(1 + \varepsilon).$$

This implies

$$\begin{aligned} J(f_1, \varphi) + J(f_2, \varphi) &\leq J(f, \varphi)(1 + \varepsilon) \\ &\leq (J(f_1 + f_2, \varphi) + \delta J(f', \varphi))(1 + \varepsilon). \end{aligned}$$

Letting δ tend to zero gives the claim. $\qquad\square$

Let F be a countable subset of $C_c^+(G)$, and let V_F be the complex vector space spanned by all translates $L_s f$, where $s \in G$ and $f \in F$. A linear functional $I : V_F \to \mathbb{C}$ is called an *invariant integral* on V_F if $I(L_s f) = I(f)$ holds for every $s \in G$ and every $f \in V_F$ and

$$f \in F \quad \Rightarrow \quad I(f) \geq 0.$$

An invariant integral I_F on V_F is called *extensible* if for every countable set $F' \subset C_c^+(G)$ that contains F there is an invariant integral $I_{F'}$ on $V_{F'}$ extending I_F.

Lemma B.5 *For every countable set $F \subset C_c^+(G)$ there exists an extensible invariant integral I_F that is unique up to scaling.*

Proof: Fix a metric on G. For $n \in \mathbb{N}$ let $\varphi_n \in C_c^+(G)$ be nonzero with support in the open ball of radius $1/n$ around the unit. Suppose that $\varphi_n(x) = \varphi_n(x^{-1})$ for every $x \in G$.

Let $F = \{f_1, f_2, \dots\}$. Since the sequence $J(f_1, \varphi_n)$ lies in the compact interval $[1/(f_0 : f_1), (f_1 : f_0)]$ there is a subsequence φ_n^1 of φ_n such that $J(f_1, \varphi_n^1)$ converges. Next there is a subsequence φ_n^2 of φ_n^1 such that $J(f_2, \varphi_n^2)$ also converges. Iterating this gives a sequence (φ_n^j) of subsequences. Let $\psi_n = \varphi_n^n$ be the diagonal sequence. Then for every $j \in \mathbb{N}$ the sequence $(J(f_j, \psi_n))$ converges, so that the definition

$$I_{f_0, (\psi_n)_{n \in \mathbb{N}}}(f_j) = \lim_{n \to \infty} J(f_j, \psi_n)$$

makes sense. By Lemma B.4, the map I indeed extends to a linear functional on V_F that clearly is a nonzero invariant integral.

This integral is extensible, since for every countable $F' \supset F$ in $C_c^+(G)$ one can iterate the process and go over to a subsequence of ψ_n.

This does not alter I_{f_0,ψ_n}, since every subsequence of a convergent sequence converges to the same limit.

We shall now establish the uniqueness. Let $I_F = I_{f_0,\psi_n}$ be the invariant integral just constructed. Let I be another extensible invariant integral on V_F. Let $f \in F$, $f \neq 0$; then we will show that

$$I_{f_0,\psi_n}(f) = \frac{I(f)}{I(f_0)}.$$

The assumption of extensibility will enter our proof in that we will freely enlarge F in the course of the proof. Now let the notation be as in the lemma. Let $\varphi \in F$ and suppose $f(x) \leq \sum_{j=1}^m d_j\varphi(s_j x)$ for some positive constants d_j and some elements s_j of G. Then

$$I(f) \leq \sum_{j=1}^m d_j I(\varphi),$$

and therefore

$$\frac{I(f)}{I(\varphi)} \leq (f : \varphi).$$

Let $\varepsilon > 0$. Since f is uniformly continuous, there is a neighborhood V of the unit such that for $x, s \in G$ we have $x \in sV \Rightarrow |f(x) - f(s)| < \varepsilon$. Let $\varphi \in C_c^+$ be zero outside V and suppose $\varphi(x) = \varphi(x^{-1})$. Let C be a countable dense set in G. The existence of such a set is clear by Lemma 6.3.1. Now suppose that for every $x \in C$ the function $s \mapsto f(s)\varphi(s^{-1}x)$ lies in F. For $x \in C$ consider

$$\int_G f(s)\varphi(s^{-1}x)ds = I(f(.)\varphi(.^{-1}x)).$$

Now, $\varphi(s^{-1}x)$ is zero unless $x \in sV$, so

$$\int_G f(s)\varphi(s^{-1}x)ds > (f(x) - \varepsilon)\int_G \varphi(s^{-1}x)ds$$
$$= (f(x) - \varepsilon)\int_G \varphi(x^{-1}s)ds$$
$$= (f(x) - \varepsilon)I(\varphi).$$

Therefore,

$$(f(x) - \varepsilon) < \frac{1}{I(\varphi)}\int_G f(s)\varphi(s^{-1}x)ds.$$

Let $\eta > 0$, and let W be a neighborhood of the unit such that

$$x, y \in G, \quad x \in Wy \Rightarrow |\varphi(x) - \varphi(y)| < \eta.$$

There are finitely many $s_j \in G$ and $h_j \in C_c^+(G)$ such that the support of h_j is contained in $s_j W$ and

$$\sum_{j=1}^{m} h_j \equiv 1 \quad \text{on supp}(f).$$

Such functions can be constructed using the metric (see Exercise 8.2). We assume that for each j the function $s \mapsto f(s)h_j(s)\varphi(s^{-1}x)$ lies in F for every $x \in C$. Then it follows that

$$\int_G f(s)\varphi(s^{-1}x)ds = \sum_{j=1}^{m} \int_G f(s)h_j(s)\varphi(s^{-1}x)ds.$$

Now $h_j(s) \neq 0$ implies $s \in s_j W$, and this implies

$$\varphi(s^{-1}x) \leq \varphi(s_j^{-1}x) + \eta.$$

Assuming that the $f h_j$ lie in F, we conclude that

$$\int_G f(s)\varphi(s^{-1}x)ds \leq \sum_{j=1}^{m} I(f h_j)(\varphi(s_j^{-1}x) + \eta).$$

Let $c_j = I(h_j f)/I(\varphi)$; then $\sum_j c_j = I(f)/I(\varphi)$ and

$$f(x) \leq \varepsilon + \eta \sum_{j=1}^{m} c_j + \sum_{j=1}^{m} c_j \varphi(s_j^{-1}x).$$

Let $\chi \in C_c^+(G)$ be such that $\chi \equiv 1$ on supp(f). Then

$$f(x) \leq \left(\varepsilon + \eta \sum_{j=1}^{m} c_j \right) \chi(x) + \sum_{j=1}^{m} c_j \varphi(s_j^{-1}x).$$

This result is valid for $x \in C$ in the first instance, but the denseness of C implies it for all $x \in G$. As $\eta \to 0$ it follows that

$$(f : \varphi) \leq \varepsilon(\chi : \varphi) + \frac{I(f)}{I(\varphi)}.$$

Therefore,

$$\frac{(f : \varphi)}{(f_0 : \varphi)} \leq \varepsilon \frac{(\chi : \varphi)}{(f_0 : \varphi)} + \frac{I(f)}{I(\varphi)(f_0 : \varphi)} \leq \varepsilon \frac{(\chi : \varphi)}{(f_0 : \varphi)} + \frac{I(f)}{I(f_0)}.$$

So, as $\varepsilon \to 0$ and as φ runs through the ψ_n, we get

$$I_{f_0, \psi_n}(f) \leq \frac{I(f)}{I(f_0)}.$$

Applying the same argument with the roles of f and f_0 interchanged gives

$$I_{f, \psi_n}(f_0) \leq \frac{I(f_0)}{I(f)}.$$

Now note that both sides of these inequalities are antisymmetric in f and f_0, so that the second inequality gives

$$I_{f_0, \psi_n}(f) = I_{f, \psi_n}(f_0)^{-1} \geq \left(\frac{I(f_0)}{I(f)}\right)^{-1} = \frac{I(f)}{I(f_0)}.$$

Thus it follows that $I_{f_0, \psi_n}(f) = I(f)/I(f_0)$ and the lemma is proven.
□

Finally, the proof of the theorem proceeds as follows. For every countable set $F \subset C_c^+(C)$ with $f_0 \in F$, let I_F be the unique extensible invariant integral on V_F with $I_F(f_0) = 1$. We define an invariant integral on all $C_c(G)$ as follows: For $f \in C_c^+(G)$ let

$$I(f) = I_{\{f_0, f\}}(f).$$

Then I is additive, since for $f, g \in C_c^+(G)$,

$$\begin{aligned}
I(f + g) &= I_{\{f_0, f+g\}}(f + g) = I_{\{f_0, f, g\}}(f + g) \\
&= I_{\{f_0, f, g\}}(f) + I_{\{f_0, f, g\}}(g) = I_{\{f_0, f\}}(f) + I_{\{f_0, g\}}(g) \\
&= I(f) + I(g).
\end{aligned}$$

Thus I extends to an invariant integral on $C_c(G)$, with the invariance being clear from Lemma B.5.
□

Bibliography

[1] **Benedetto, J.**: *Harmonic Analysis and Applications.* Boca Raton: CRC Press 1997.

[2] **Bröcker, T.; tom Dieck, T.**: *Representations of Compact Lie Groups.* Springer-Verlag 1985.

[3] **Dixmier, J.**: *Les C* algèbres et leur représentations.* (2nd ed.) Gauthier Villars, Paris 1969.

[4] **Edwards, R. E.**: *Fourier Series. A Modern Introduction.* Springer-Verlag 1979 (vol I) 1981 (vol II).

[5] **Khavin, V.P. (ed.); Nikol'skij, N.K. (ed.); Gamkrelidze, R.V. (ed.)**: *Commutative harmonic analysis I. General survey. Classical aspects.* Encyclopaedia of Mathematical Sciences, 15. Berlin etc.: Springer-Verlag. 1991.

[6] **Havin, V.P. (ed.); Nikolski, N.K. (ed.): Gamkrelidze, R.V. (ed.)**: *Commutative harmonic analysis II. Group methods in commutative harmonic analysis.* Encyclopaedia of Mathematical Sciences. 25. Berlin: Springer-Verlag. 1998.

[7] **Havin, V.P. (ed.); Nikol'skij, N.K. (ed.); Gamkrelidze, R.V. (ed.)**: *Commutative harmonic analysis III. Generalized functions. Applications.* Encyclopaedia of Mathematical Sciences. 72. Berlin: Springer-Verlag. 1995.

[8] **Khavin, V.P. (ed.); Nikol'skij, N.K. (ed.); Gamkrelidze, R.V. (ed.)**: *Commutative harmonic analysis IV: harmonic analysis in R^n.* Encyclopaedia of Mathematical Sciences. 42. Berlin etc.: Springer-Verlag. 1992.

[9] **Helgason, S.**: *Differential Geometry, Lie Groups, and Symmetric Spaces.* Academic Press 1978.

[10] **Helson, H.:** *Harmonic analysis.* 2nd ed. Texts and Readings in Mathematics. 7. New Delhi: Hindustan Book Agency. 1995.

[11] **Hewitt, E.; Ross, K.A.:** *Abstract harmonic analysis. Volume I: Structure of topological groups, integration theory, group representations.* Grundlehren der Mathematischen Wissenschaften. 115. Berlin: Springer-Verlag 1994.

[12] **Hewitt, Edwin; Ross, Kenneth A.:** *Abstract harmonic analysis. Vol. II: Structure and analysis for compact groups. Analysis on locally compact Abelian groups.* Grundlehren der Mathematischen Wissenschaften. 152. Berlin: Springer-Verlag 1994.

[13] **Karatsuba, A.:** *Basic Analytic Number Theory.* Springer-Verlag. 1993.

[14] **Knapp, A.:** *Representation Theory of Semisimple Lie Groups.* Princeton University Press 1986.

[15] **Körner, T.:** *Exercises for Fourier Analysis.* Cambridge University Press 1993.

[16] **Krantz, S.:** *A panorama of harmonic analysis.* The Carus Mathematical Monographs. 27. Washington, DC: The Mathematical Association of America. 1999.

[17] **Lang, S.:** *Algebra.* 3rd ed. Addison Wesley. 1993.

[18] **Loomis, L.H.:** *Harmonic analysis.* Buffalo: Mathematical Association of America. 1965.

[19] **Seeley, R.:** *An Introduction to Fourier Series and Integrals.* New York: Benjamin 1966.

[20] **Rudin, W.:** *Real and complex analysis.* 3rd ed. New York, NY: McGraw-Hill. 1987.

[21] **Rudin, W.:** *Functional Analysis.* 2nd ed. McGraw-Hill 1991.

[22] **Stein, E. M.:** *Harmonic analysis: real-variable methods, orthogonality, and oscillatory integrals.* Princeton University Press, Princeton, NJ, 1993.

[23] **Taylor, M.:** *Noncommutative Harmonic Analysis.* AMS 1985.

[24] **Warner, F.:** *Foundations of Differentiable Manifolds and Lie Groups.* Springer-Verlag 1983.

[25] **Yosida, K.:** *Functional Analysis.* Berlin, Heidelberg, New York: Springer-Verlag 1980.

Index

190